INSTRUCTOR'S RESOURCE GUIDE

BASIC MATHEMATICS

SEVENTH EDITION

INSTRUCTOR'S RESOURCE GUIDE

BASIC MATHEMATICS

SEVENTH EDITION

Marvin L. Bittinger
Indiana University—Purdue University at Indianapolis

Mervin L. Keedy
Purdue University

Donna DeSpain
Illinois Benedictine College

ADDISON-WESLEY PUBLISHING COMPANY
Reading, Massachusetts • Menlo Park, California • New York
Don Mills, Ontario • Wokingham, England • Amsterdam • Bonn
Sydney • Singapore • Tokyo • Madrid • San Juan • Milan • Paris

Reproduced by Addison-Wesley from camera-ready copy supplied by the authors.

Copyright © 1995 Addison-Wesley Publishing Company, Inc.

ISBN 0-201-54251-X

1 2 3 4 5 6 7 8 9 10 CRS 97969594

TABLE OF CONTENTS

I. EXTRA PRACTICE SHEETS 1

These sheets provide extra drill on topics in the text. The instructor can use them for lecture examples or additional homework assignments. Students will find them to be an excellent review for tests. They provide an excellent source of practice and reteaching for students who have done poorly on a test and who are going to retest.

These essays were written by instructors from various schools around the country. They describe how the mathematics lab works, including what types of courses use the lab, the type and number of students, the level of mastery required, how tutors interact with students, and backgrounds of the tutors. The instructors also talk about the types of hardware and software they use, how tests are generated and administered, and major challenges within their developmental program.

A list of these lab coordinators has been included to provide a resource for you. If you have further questions, please feel free to contact these people.

INSTRUCTOR'S RESOURCE GUIDE

BASIC MATHEMATICS
SEVENTH EDITION

Addition and Subtraction of Whole Numbers
Use after Sections 1.2 and 1.3   Name _____

Add.

1.	4	2.	8	3.	2	4.	7	5.	12
	9		1		18		19		36
	15		12		17		8		14
	6		6		4		23		5
	2		11		3		11		7
	+ 3		+ 9		+ 11		+ 6		+ 28

6.	57	7.	16	8.	83	9.	459	10.	273
	142		351		172		115		46
	31		229		95		356		311
	+ 265		+ 593		+ 614		+ 542		+ 986

11.	152	12.	78	13.	4352	14.	53271	15.	96488
	3649		1349		1679		165333		352567
	6517		452		35341		495672		431
	+ 836		+ 2415		+ 62487		+ 95126		+ 768259

Subtract.

16.	564	17.	963	18.	378	19.	531	20.	6611
	- 321		- 452		- 139		- 67		- 4332

21.	9534	22.	8056	23.	6503	24.	43512	25.	59384
	- 8826		- 3509		- 6417		- 29608		- 39565

26.	65003	27.	72500	28.	95036	29.	145379	30.	654042
	- 49021		- 83095		- 30427		- 95084		- 293536

Multiplication and Division of Whole Numbers
Use after Sections 1.5 and 1.6   Name _____

Multiply.

1.   62
   × 27

2.   55
   × 21

3.   147
   × 23

4.   327
   × 29

5.   456
   × 661

6.   977
   × 649

7.   473
   × 725

8.   1542
   × 333

9.   5630
   × 4222

10.   7459
   × 5099

11.   6721
   × 942

12.   9655
   × 8176

13.   5278
   ×   29

14.   41699
   ×   122

15.   256
   × 38

16.   2906
   ×   54

Divide and check.

17. 8)6890

18. 21)55

19. 35)79

20. 56)242

21. 47)546

22. 26)589

23. 59)3586

24. 129)561

25. 283)894

26. 748)7037

27. 788)5644

28. 916)9056

29. 2236)9692

30. 4921)5237

31. 3162)11971

32. 4938)54563

Solve. (Be careful! Some problems have numbers included which will not be used in finding the solution.)

1. The Declaration of Independence was signed in 1776, the Constitution was framed in 1787, and the national anthem was designated in 1931. How many years elapsed between the Declaration of Independence and the Constitution? _____

2. A small business sold for $140,000. The buyer had $40,000, the seller loaned $20,000, and one of the business's suppliers loaned $20,500. The remainder was borrowed from a bank. How much was the bank loan?

_____

3. The area of Alaska is 586,400 square miles. The area of Rhode Island is 1214 square miles. About how many Rhode Islands would fit into Alaska?

_____

4. How many calories is the following meal? Fried beef liver at 389 calories, a baked potato at 93 calories, 2 helpings of zucchini squash at 25 calories each, cranberry sauce at 96 calories, 4 oatmeal cookies at 60 calories each, and 2 glasses of skim milk at 81 calories each.

_____

5. The area in the United States where New Mexico, Arizona, Utah, and Colorado meet is called the Four Corners area. The areas of these four states are 121,666, 113,909, 84,916, and 104,247 square miles, respectively. The area of Texas is 267,339 square miles. How much smaller is Texas than the Four Corners states? _____

6. One package of notebook paper has 350 sheets and costs $1.57. How many sheets of paper are there in 3 packages? _____

7. During the years of 1891-1900, there were 14,799 immigrants to the United States from China, 68 from India, 25,942 from Japan, 26,799 from Asian Turkey, and 3628 from other parts of Asia. How many immigrants were there altogether from Asia? _____

8. The interior noise levels of sports sedans were measured at 30 mph. The lowest level was 63 decibels and the highest level was 70 decibels. How much more was the highest than the lowest? _____

9. A man bought 4 tires at $33 each, and paid with $20 bills. How many $20 bills did it take, and how much change did he receive, if the sales tax was $8? _____

10. A 4-pound bag of sugar is 64 ounces. How many 3-ounce sugar bowls can be filled from one 4-pound bag, and how many ounces will be left over?

_____

11. Minneapolis and St. Paul have populations of 370,951 and 270,230 people, respectively. Dallas and Fort Worth have populations of 904,078 and 385,141 people, respectively. How many more people live in the Dallas-Fort Worth area than in the Twin Cities area? _____

12. Las Cruces, New Mexico, has 47,395 people. Albuquerque has 331,767 people. About how many times as large as Las Cruces is Albuquerque? _____

13. A woman who weighed 208 pounds went on a diet and now weighs 124 pounds. She took 14 months to lose the weight. How much, on the average, did she lose each month? _____

14. The Great Wall of China, parts of which still stand, was being built in the year 214 B.C. How many years ago was that? Use the year 1994 for the present. _____

15. A cookie recipe makes 3 dozen cookies. How many times is the recipe needed for 144 people, if each person has 2 cookies? _____

16. The average distance of the earth from the sun is 92,956,000 miles. The average distance of Mars from the sun is 141,637,000 miles. About how far apart are the earth and Mars? _____

17. A 24-inch, 50-pound attic fan costs $130. Shipping is $24 for each 100 pounds. How much will it cost a building contractor to buy 10 fans?

_____

18. A 9-inch pie serves 6 people and a 10-inch pie serves 7 people. How many people will 3 of the larger and 4 of the smaller pie serve? _____

19. A patient has been instructed to take pills, each of which has 60 mg of a drug. Two pills are to be taken in the morning, one is to be taken at 3 pm, and two at bedtime. How many days will 100 pills last? _____

Name _____

Simplify using the rules for order of operations.

1. $5 + 10 - 3^2$ _____

2. $6 \times 4 + 5^2 - 11$ _____

3. $7 \times (2 + 3) - 21$ _____

4. $(5 + 7) \div 2^2$ _____

5. $5^2 - 4^2 + 3 \times 2$ _____

6. $8 \times 9 - 6^2 \div 4$ _____

7. $9^2 - (20 + 11)$ _____

8. $(2 + 3) \times 10^2 \div 5^2$ _____

9. $3 \times (30 + 4) - 7^2$ _____

10. $(1 + 5) \times 5 - 7 \times 4$ _____

11. $0 \times 15^2 \times (400 + 21) \div 19^2$ _____

12. $0 \times 17^2 \times (59 + 92) + 5$ _____

13. $6 \times (5 + 0)^2$ _____

14. $(7 - 7) \times 33^2 \div (45 + 3)^2$ _____

15. $49 \div 7^2 \times (531 + 4)$ _____

16. $(233 + 17) \div 250 + 3 \times 33$ _____

17. $\dfrac{5 \times 30}{15} - (3 + 3)$ _____

18. $\dfrac{(10 + 14) \times 200}{10^2}$ _____

19. $\dfrac{10 \times (25 + 7)}{(5 + 3)^2}$ _____

20. $\dfrac{25 \times (6 + 7) - 5^2}{(6 + 7)^2 - 19}$ _____

21. $3 \times 4 \div (1 + 2) + 5$ _____

22. $(9 + 6) \times 3 \div (18 - 13)$ _____

23. $(1^6 + 13) \times 2 \div 7$ _____

24. $(8^2 - 2^2) + 80$ _____

EXAMPLE: Give the factors and multiples of 18.
Factors: 1, 2, 3, 6, 9, 18
Multiples: 18, 36, 54, 72, . . .

Give the factors and the first four multiples of each of the following
numbers.

1. 6; _____   2. 12; _____

_____   _____

3. 20; _____   4. 45; _____

_____   _____

5. 50; _____   6. 21; _____

_____   _____

7. 18; _____   8. 30; _____

_____   _____

9. 27; _____   10. 35; _____

_____   _____

11. 56: _____   12. 39; _____

_____   _____

13. 80; _____   14. 37; _____

_____   _____

15. 55; _____   16. 53; _____

_____   _____

17. 100; _____ 18. 125; _____

_____   _____

19. 128; _____ 20. 111; _____

_____   _____

Tests for divisibility.

Divisibility by 2:   any even number.
                3:   sum of digits is divisible by 3.
                4:   last two digits are divisible by 4.
                5:   ones digit is 0 or 5.
                6:   divisible by both 2 and 3.
                7:   actually divide.
                8:   last three digits are divisible by 8.
                9:   sum of digits is divisible by 9.
               10:   ones digit is 0.

Using the tests for divisibility, determine which of the numbers 2, 3, 4, 5, 6, 7, 8, 9, 10, divide the following.

1. 378 is divisible by _____   2. 3135 is divisible by _____

3. 550 is divisible by _____   4. 625 is divisible by _____

5. 572 is divisible by _____   6. 1001 is divisible by _____

7. 280 is divisible by _____   8. 540 is divisible by _____

9. 612 is divisible by _____   10. 1848 is divisible by _____

11. 2025 is divisible by _____   12. 513 is divisible by _____

13. 1320 is divisible by _____   14. 372 is divisible by _____

15. 138 is divisible by _____   16. 128 is divisible by _____

17. 3000 is divisible by _____   18. 8550 is divisible by _____

19. 139 is divisible by _____   20. 146 is divisible by _____

Name _____

EXAMPLE: Find the prime factorization of 140.

a) We divide by the first prime, 2. $140 = 2 \cdot 70$

b) We continue to divide by primes. $140 = 2 \cdot 2 \cdot 35$

$140 = 2 \cdot 2 \cdot 5 \cdot 7$

c) We are finished when we have a product of primes.   The prime factorization is $140 = 2 \cdot 2 \cdot 5 \cdot 7$

Find the prime factorization.

1. 378 = _____

2. 3135 = _____

3. 550 = _____

4. 625 = _____

5. 572 = _____

6. 1001 = _____

7. 280 = _____

8. 540 = _____

9. 612 = _____

10. 1848 = _____

11. 2025 = _____

12. 513 = _____

13. 1320 = _____

14. 372 = _____

15. 138 = _____

16. 128 = _____

17. 3000 = _____

18. 8550 = _____

19. 139 = _____

20. 146 = _____

Operations with Fractional Notation
Use after Sections 2.6 - 3.3   Name _____

Perform the indicated operations.   Simplify if possible.

1. $\frac{3}{4} \cdot \frac{7}{8} =$ _____

2. $\frac{1}{2} \div \frac{1}{4} =$ _____

3. $\frac{3}{5} + \frac{4}{5} =$ _____

4. $\frac{15}{8} - \frac{5}{8} =$ _____

5. $\frac{5}{2} \div \frac{3}{8} =$ _____

6. $5 \cdot \frac{4}{7} =$ _____

7. $\frac{4}{3} + \frac{1}{2} =$ _____

8. $\frac{3}{7} \div 4 =$ _____

9. $\frac{9}{11} - \frac{1}{3} =$ _____

10. $\frac{8}{7} \cdot \frac{21}{16} =$ _____

11. $\frac{7}{5} - \frac{4}{3} =$ _____

12. $\frac{4}{9} + \frac{6}{27} =$ _____

13. $5 \div \frac{5}{13} =$ _____

14. $\frac{5}{13} \div 5 =$ _____

15. $\frac{9}{10} + \frac{3}{7} =$ _____

16. $\frac{13}{16} - \frac{5}{8} =$ _____

17. $\frac{11}{15} \times \frac{5}{22} =$ _____

18. $\frac{26}{20} - \frac{2}{3} =$ _____

19. $\frac{3}{5} \div \frac{9}{10} =$ _____

20. $4 + \frac{3}{7} =$ _____

21. $\frac{11}{25} + \frac{3}{4} =$ _____

22. $\frac{15}{24} \times \frac{6}{25} =$ _____

23. $\frac{5}{4} \cdot 16 =$ _____

24. $\frac{1}{6} + \frac{5}{8} =$ _____

25. $\frac{8}{15} - \frac{2}{10} =$ _____

26. $\frac{6}{7} \div 6 =$ _____

27. $6 \div \frac{6}{7} =$ _____

28. $\frac{1}{9} \cdot \frac{9}{10} =$ _____

29. $\frac{1}{6} \cdot \frac{1}{8} =$ _____

30. $\frac{2}{5} + 10 =$ _____

EXTRA PRACTICE 9
Operations with Mixed Numerals
Use after Sections 3.5 and 3.6   Name _____

1. $5 + 4\frac{5}{7} =$ _____

2. $1\frac{2}{5} \times 2\frac{1}{3} =$ _____

3. $2\frac{1}{4} + 6\frac{1}{4} =$ _____

4. $3\frac{1}{2} \div 2 =$ _____

5. $6\frac{4}{5} - 3\frac{2}{3} =$ _____

6. $5\frac{3}{4} \cdot 1\frac{3}{5} =$ _____

7. $10\frac{1}{6} + 3\frac{2}{3} =$ _____

8. $12 \div 1\frac{1}{2} =$ _____

9. $9\frac{2}{7} - 4\frac{4}{7} =$ _____

10. $8\frac{4}{5} + 1\frac{2}{5} =$ _____

11. $3\frac{1}{4} \cdot 5\frac{3}{4} =$ _____

12. $6\frac{7}{11} + 5\frac{4}{11} =$ _____

13. $11\frac{5}{8} - 5\frac{5}{8} =$ _____

14. $10\frac{5}{6} - 4\frac{3}{4} =$ _____

15. $13\frac{1}{3} - 7\frac{3}{4} =$ _____

16. $8 - 3\frac{13}{16} =$ _____

17. $12\frac{5}{8} + 4\frac{3}{4} =$ _____

18. $6\frac{2}{3} \cdot 5\frac{1}{4} =$ _____

19. $9\frac{3}{8} \div 1\frac{5}{6} =$ _____

20. $5\frac{1}{2} \div 5\frac{1}{2} =$ _____

21. $0 \times 76\frac{5}{9} =$ _____

22. $26\frac{1}{2} + 14\frac{7}{8} =$ _____

23. $45\frac{1}{6} - 8\frac{5}{8} =$ _____

24. $58\frac{4}{7} - 10 =$ _____

25. $67 - 5\frac{4}{5} =$ _____

26. $35\frac{2}{7} - 21\frac{2}{3} =$ _____

27. $0 \div 65\frac{2}{3} =$ _____

28. $7\frac{5}{6} \div 10 =$ _____

29. $75\frac{2}{3} - 48\frac{7}{9} =$ _____

30. $15\frac{5}{6} \cdot 8\frac{1}{10} =$ _____

Solve. Be sure to read carefully!

1. A waffle recipe says to use $1\frac{3}{4}$ c pancake mix, $1\frac{1}{4}$ c cold water, 1 egg, $\frac{1}{2}$ c bran cereal, and 2 T oil. How much of each ingredient is needed for $\frac{1}{3}$ of the recipe? How could you measure the amounts of pancake mix, water, and egg? _____

2. A shoo-fly pie is cut into 6 pieces. A tamale pie of the same size is cut into 5 pieces. The pieces of which pie are larger? Which fraction is larger, $\frac{1}{6}$ or $\frac{1}{5}$? _____

3. A baker bought a 50-lb bag of flour. He used $\frac{1}{3}$ of it one day, $\frac{1}{6}$ the next day, and $\frac{1}{4}$ the third day. What fraction of the bag was left? How many pounds were left? _____

4. A car measures $\frac{3}{1000}$ mi long. The distance from Los Angeles to San Francisco is 425 miles. About how many cars could be lined up end to end between the two cities? _____

5. A car was bought for $8900. The down payment was $\frac{1}{5}$ of the price, the trade-in amounted to $\frac{1}{10}$ of the price, and the rest was borrowed. How much money was borrowed? _____

6. The price of stock was $\$28\frac{1}{8}$. The price rose $\$\frac{1}{4}$, and then declined $\$\frac{5}{8}$. What was the resulting price? _____

7. A pipe $\frac{7}{8}$ yd long is cut into 3 pieces. How long is each piece? _____

8. From a board $5\frac{1}{2}$ ft long, 3 pieces are to be cut. One is to be $1\frac{2}{3}$ ft, the second $1\frac{1}{4}$ ft, and the third $2\frac{1}{2}$ ft. How long will the leftover piece be? _____

9. A stretch of highway is $28\frac{1}{3}$ mi long. Already $\frac{2}{5}$ of it has been repaved. How many miles still need to be repaved? _____

10. A box of cold cereal says that one serving of the cereal with milk provides 6 grams of protein, which is $\frac{3}{20}$ (15%) of the U.S. recommended daily allowance (RDA) for protein. What is the U.S. RDA for protein? _____

11. An area of 640 sq ft is being painted. The painter has already finished 240 sq ft. What fraction of the area is still to be painted? _____

12. A package of hamburger meat weighs $\frac{3}{4}$ lb. If a person eats half of the package, how much of a pound is eaten? Is the amount more or less than a quarter-pounder? _____

13. A container holds $1\frac{1}{2}$ gal of lemonade. How many people can be served if each person drinks $\frac{1}{6}$ gal? _____

14. A full-time salary is \$19,200. What would a person's salary be working $\frac{2}{3}$-time? _____

15. A man's hourly wage is \$6 for 40 hours. He gets $1\frac{1}{2}$ times that for work over 40 hours. How much is his paycheck for a 45-hour week? _____

16. A bookworm is eating its way through 2 books, each with pages $1\frac{1}{8}$ in. thick when closed. The covers of each book are each $\frac{3}{16}$ in. thick. How far has the bookworm traveled when it has gone completely through the 2 books?

    _____

17. A roll of plastic sheeting is 25 yd long. How many pieces can be cut from the roll, if each piece is $1\frac{1}{4}$ yd long? _____

18. A mechanic picks up a $\frac{1}{2}$-in. socket and a $\frac{5}{8}$-in. socket. One is too large and the other too small. Her set of sockets does have sixteenths of an inch. Which one should she try? By how much are the others too large and too small? _____

19. A Christmas tree farm is 20 acres. The owners want $6\frac{1}{2}$ acres planted with long-needle pines, $3\frac{1}{4}$ acres of pinon pine, and $5\frac{3}{4}$ with fir trees. The rest are to be planted with spruce trees. How many acres are left for spruce trees? _____

20. A home economics teacher wants to give each of 21 students a piece of material cut from a bolt which has $18\frac{2}{3}$ yd left on it. Should he give them a little less than a yard each, or a little more? Exactly how long would each piece be to use the whole bolt? _____

Solve. (Be careful! Some problems have numbers included which will not be used in finding the solution.)

1. Joe Blow has $531.67 in his checking account, account number 643092. He deposits his paycheck of $456.19 and another check for $24. He writes 2 checks, one for $36.55, and the other for $73.14. How much is his new checkbook balance? _____

2. A checking account balance shows $652.35. Check number 387 is written to buy 4 tires at $29.99 each, plus tax of $6.30. What is the new balance?

   _____

3. A checker board is 506.25 square inches. How many square inches is each of the 64 squares? Round to the nearest tenth. _____

4. A driver bought 15.3 gallons of gasoline when the odometer read 25623.4. She bought 10.8 gallons of gasoline when the odometer read 25856.8. How many miles were driven? _____

5. When filling the gasoline tank a driver noted that the odometer read 43561.2. The next time he bought gasoline, the tank took 12.7 gallons, and the odometer read 43755.5. How many miles per gallon did the car average? Round to the nearest tenth. _____

6. Postage stamps cost 25¢ each. How much will 70 stamps cost? _____

7. Stamps for postcards cost 15¢ each. How many can be bought for $6.00?

   _____

8. The United States flag has 7 red stripes and 6 white stripes. Each stripe on a particular flag is 3.5 inches wide. How wide is the flag? _____

9. A seamstress needs 1.4 meters of cloth for a blouse, 2.5 meters for a skirt, and 2.2 meters for a jacket. She has 6.3 meters of cloth. Is this enough? How much more will be needed, or how much will be left over?

   _____

10. One cup of $1\frac{1}{2}$% butterfat milk has 0.291 g of calcium. How many grams of calcium do 4.5 cups of the same milk have? Round to the nearest hundredth. _____

11. A package of 3 bars of soap costs $0.71. How many bars can be bought for $4.00? _____

12. Marcy spent $1.49 for a hamburger, 89¢ for a milkshake, and 33¢ for a pickle. Belinda spent $1.99 for a chicken sandwich and 79¢ for a cherry pie. Who spent more for lunch? How much more? _____

13. Todd babysits his 2 neighbor children for $2.25 an hour. One week he babysat 4 hours one day, 3 hours the next, and 5 hours the third day. How much money did he earn? _____

14. In a driving test, Car #1 had an average stopping distance of 123.60 ft (going from 50 mph to a complete stop). Car #2 had an average stopping distance of 124.20 ft. How many more feet did Car #2 need to stop?

_____

15. A textbook costs $22.95. How many can be bought for $1000? Each book has 368 pages. _____

16. Vacation meals, per person, cost $2.50 for breakfast, $3.00 for lunch, and $6.75 for dinner. How much will the meals for a full week cost for 2 people? _____

17. An 8-inch computer disk costing $4.25 holds 1.25 megabytes. How many megabytes do 6 disks have? _____

18. Each board of a fence is 3.5 m long. How many boards will be needed for a 770 meter-long fence, if the fence is to be 4 boards high? (It might help to draw a picture.) _____

19. The length of a bacterium is 0.00003 cm. How many of this bacterium will fit length-to-length across a tooth that is 0.75 cm wide? _____

20. Using a vernier caliper, a machinist measures the diameter of a 10-inch rod to be 0.762 in. She needs a rod with a 0.75 in. diameter. How much of an inch does she need to machine off? _____

14

Solve using proportions. (Be careful! Some of the problems include numbers which will not be needed in finding the solution.)

1. A family of 5 travels 1230 miles in 4 days. At the same rate, how long would the family take to travel 5535 miles? _____

2. A 1987 automobile goes 229.5 miles on 9 gallons of gasoline. At the same rate, how far would the automobile go on 20 gallons of gasoline? _____

3. For a school picnic, 2 pounds of 85% lean hamburger meat serves 9 people. How many pounds of the same hamburger would be needed to serve 170 people? Round to the nearest whole pound. _____

4. A 6-cylinder car is driven 37,500 miles in 3 years. At that rate, in how many years will the car have 100,000 miles on it? _____

5. On a 50-pound bag of fertilizer, the directions say to apply 2.5 lb of fertilizer for every 100 square feet 4 times a year. How many pounds of fertilizer are needed for 2500 square feet each time? _____

6. On a scale model of an airplane, $\frac{1}{2}$ in. represents 2 ft. The wingspan on the actual airplane is 44 ft. How long will the model's wingspan be?

_____

7. A punch recipe which serves 12 people uses 1 cup of orange juice and $1\frac{1}{4}$ cups of pineapple juice. How much pineapple juice will be needed to serve 72 people? _____

8. 1 foot is 12 inches. How many feet is 102 inches? _____

9. 36 inches is 3 feet. How many inches is 13 feet? _____

10. One pound is 16 ounces. How much of a pound is 12 ounces? _____

11. There are 36 barleycorns (a barleycorn is an old unit of measure) in 12 inches. How many barleycorns are there in 8 inches? _____

12. One inch is 2.54 centimeters. How many centimeters is 12 inches? _____

13. One kilometer is 5/8 mile. How many kilometers is 55 miles? _____

14. The directions on a box of food coloring say, to obtain the color orange, mix 2 parts of red coloring to 3 parts of yellow. You have already used 8 drops of red. How many drops of yellow are needed? _____

15. $900 in the bank earns $76.50 in simple interest. How much would $400 earn in the same amount of time at the same interest rate? _____

16. The rectangles below have the same shape, so the lengths of their sides are proportional. Find the length of the unknown side. _____

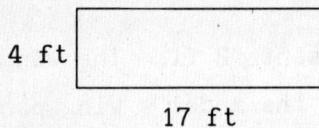

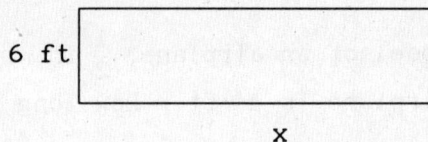

4 ft [   ]   6 ft [   ]
      17 ft   x

17. Corresponding sides of the right triangles below are proportional. Find the length of the unknown side. _____

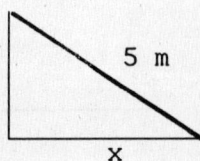

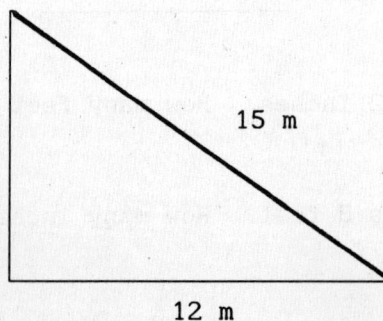

5 m
  x

15 m

12 m

EXTRA PRACTICE 13
Solving Problems Involving Percents
Use after Sections 7.3 - 7.8   Name _____

Solve.

1. A person's taxable income is $18,500 one year. Federal income tax is 15% of that income. How much would that person pay for federal income tax?

   _____

2. A drug company is having a $33\frac{1}{3}$% more sale. Normally a 9-ounce bottle of lotion sells for $1.99. With the sale, the same price buys $33\frac{1}{3}$% more. How much lotion can be bought for $1.99? _____ (Remember that $33\frac{1}{3}$% = $\frac{1}{3}$. The fraction will be easier to use.)

3. A baseball player has a batting average of 0.296, or 29.6%. The player has been at bat 135 times. How many hits did he have? _____

4. A real estate agent receives a $3850 commission on the sale of a $55,000 home. What was her rate of commission? _____

5. The risk of developing the most dangerous form of skin cancer has increased 900% since 1930. There were 130 cases of this skin cancer in 1930. How many cases are expected now? _____

6. After making a payment, a person is left with a credit card balance of $315. How much interest will be charged on that amount next month if the interest rate is 1.5% per month? Round to the nearest cent. _____

7. A package of hamburger meat is 70% lean meat. What percent is fat?

   _____

8. A purchase of $44.00 has a sales tax of $2.31. What is the sales tax rate? _____

9. Calcium carbonate is 40% calcium. A person takes 500 mg of calcium carbonate as a diet supplement. How much actual calcium does the person receive? _____

10. A microwave oven is on sale for $169.99. The regular price was $229.99. What percent discount is the sale price? Round to the nearest whole percent. _____

11. A pair of jeans that regularly costs $26 is on sale at 40% off. What is the sale price? _____

12. It is estimated that 12.5% of workers in the United States use a computer in their job. If 15 million workers use a computer, how many workers are there in total? _____

13. In the previous problem, of the 15 million workers in the United States who use a computer in their job, only 5% require extensive education in computers. How many require extensive education? _____

14. Interest earned by a bond is 8%, compounded semiannually. How much will a $750 bond be worth in one year? _____

15. After tax reform, it is estimated that a person will have a tax increase of 0.9%. How much more will that person pay in taxes, if his current tax bill is $3130? Round to the nearest dollar. _____
How much will the next total tax bill be? _____

16. A school system is required to cut its budget by 2%. Its current budget is $135 million. What will be its new budget? _____

17. The current recommended daily allowance (U.S. RDA) of vitamin C is 60 mg. A person takes 360 mg of vitamin C. What percent of the RDA is the person getting? _____

18. What is the simple interest on a loan of $8000 at 10.5% for 3 years?

_____

EXTRA PRACTICE 14
Averages, Medians, and Modes
Use after Section 8.1

Name _____

EXAMPLE: Find the average, the median, and the mode of the set of numbers:
4, 9, 7, 6, 5, 5

Average: $\frac{4 + 9 + 7 + 6 + 5 + 5}{6} = \frac{36}{6} = 6$

Median (middle score): 4, 5, ⌊5, 6,⌋ 7, 9 $\frac{5 + 6}{2} = 5.5$
↑

Mode (number which occurs most often): 5

For each set of numbers, find the average, the median, and the mode.

1. 5, 5, 9, 8, 6, 8, 8

Average _____

Median _____

Mode _____

2. 25, 32, 47, 56

Average _____

Median _____

Mode _____

3. 1.3, 4.7, 2.6, 3.8, 2.6

Average _____

Median _____

Mode _____

4. 343, 392, 371, 371, 382

Average _____

Median _____

Mode _____

5. 29, 29, 29, 27, 25, 23

Average _____

Median _____

Mode _____

6. $36.25, $48.36, $48.36, $27.31

Average _____

Median _____

Mode _____

7. 3, 6, 9, 12, 15, 18

Average _____

Median _____

Mode _____

8. 79, 64, 69, 79, 84

Average _____

Median _____

Mode _____

9. A college basketball team scored 85, 93, 80, 85, 92, 91, and 90 in its first seven games. What was the average score? the median? the mode?

Average _____

Median _____

Mode _____

10. The high temperatures for seven days were: 79°, 83°, 88°, 73°, 75°, 75°, 80°. What was the average temperature? the median? the mode?

Average _____

Median _____

Mode _____

11. To get an A in a statistics course a student must score an average of 90 on four exams. The scores on the first three exams were 87, 95, and 92. What is the lowest score that the student can get on the last test and still get an A? _____

12. To pass a physics course a student must score an average of 60 on five exams. The scores on the first four exams were 55, 67, 62, and 70. What is the lowest score the student can get on the last test and still pass the course? _____

13. The student in Exercise 12 must score an average of 70 to get a C in the course. What is the lowest score that the student can get on the last test and still get a C? _____

In Exercises 14 and 15 are the grades of a student for one semester. In each case find the grade point average. Assume that the grade point values are 4.00 for an A, 3.00 for a B, and so on.

14. Grades   Number of hours in course

A	2
B	4
B	4
C	5

_____

15. Grades   Number of hours in course

B	5
B	4
C	5
C	4

_____

Name _____

Complete.

1. 15 ft = _____ in.　　2. 37 ft = _____ in.　　3. 43 yd = _____ ft

4. 126 yd = _____ ft　　5. 126 ft = _____ yd　　6. 39 in. = _____ ft

7. 4 mi = _____ ft　　8. 6 yd = _____ in.　　9. 10,560 ft = ___ mi

10. 9 in. = _____ ft　　11. 2 ft = _____ yd　　12. 5 mi = _____ yd

13. 9 in. = _____ yd　　14. 30 in. = _____ ft　　15. 14 ft = _____ yd

16. 10 mi = _____ ft　　17. 0.1 mi = _____ ft　　18. $\frac{1}{2}$ mi = _____ ft

19. $\frac{3}{4}$ yd = _____ ft　　20. 0.25 ft = _____ in.　　21. 1760 ft = _____ mi

22. 1000 yd = _____ ft　　23. 300 ft = _____ yd　　24. 6 in. = _____ yd

25. 4 in. = _____ ft　　26. 72 in. = _____ yd　　27. $1\frac{1}{2}$ ft = _____ yd

28. 36 yd = _____ ft　　29. 18,480 ft = ___ mi　　30. 37 ft = _____ yd

Complete. Remember for conversions in the metric system you need only move decimal points.

1. 5.3 km = _____ m   2. 0.6 km = _____ m   3. 30 m = _____ cm

4. 8.9 cm = _____ mm   5. 567 mm = _____ cm   6. 30 m = _____ km

7. 152 mm = _____ cm   8. 675 m = _____ km   9. 38 m = _____ cm

10. 10 cm = _____ m   11. 15.25 m = _____ cm   12. 98 km = _____ m

13. 19427 m = _____ km   14. 41 m = _____ km   15. 62 mm = _____ cm

16. 8 mm = _____ cm   17. 459 m = _____ cm   18. 45.9 m = _____ km

19. 77.5 cm = _____ mm   20. 0.6 cm = _____ mm   21. 0.007 m = _____ cm

22. 5 mm = _____ cm   23. 10 m = _____ km   24. 520 m = _____ km

25. 0.42 cm = _____ mm   26. 0.5 cm = _____ m   27. 0.25 km = _____ m

28. 5280 mm = _____ cm   29. 16417 m = _____ km   30. 9423 cm = _____ m

Name _____

Find the area. Remember to include the units in your answer.

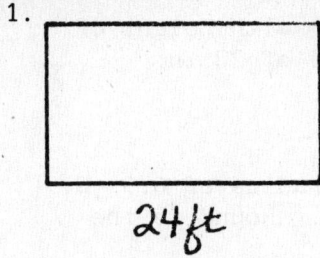

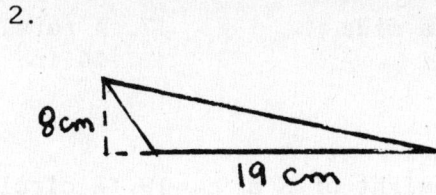

1.

13 ft

24 ft

2.

8 cm

19 cm

3. Use 3.14 for π.

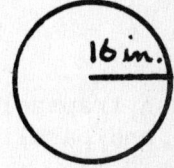

16 in.

4. Use $\frac{22}{7}$ for π.

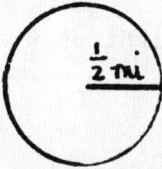

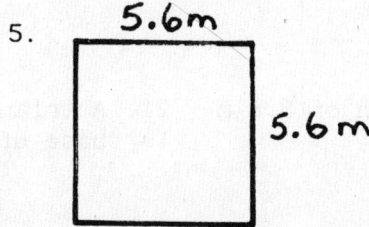

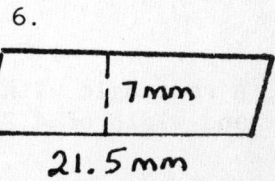

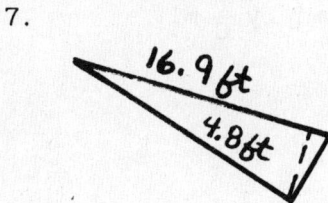

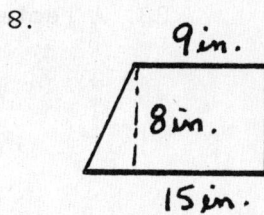

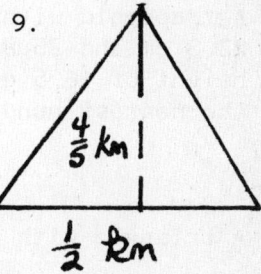

$\frac{1}{2}$ mi

5.

5.6 m

5.6 m

6.

7 mm

21.5 mm

7.

16.9 ft

4.8 ft

8.

9 in.

8 in.

15 in.

9.

$\frac{4}{5}$ km

$\frac{1}{2}$ km

10. Use 3.14 for π.
Round to the nearest
tenth.

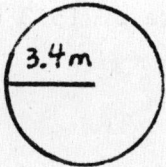

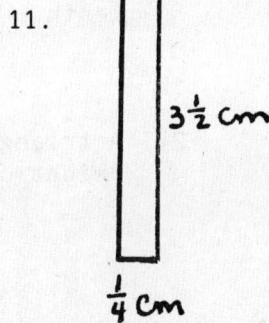

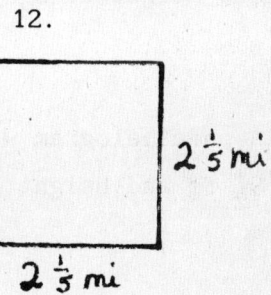

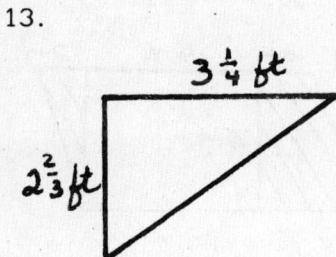

3.4 m

11.

$3\frac{1}{2}$ cm

$\frac{1}{4}$ cm

12.

$2\frac{1}{5}$ mi

$2\frac{1}{5}$ mi

13.

$3\frac{1}{4}$ ft

$2\frac{2}{3}$ ft

14. Use $\frac{22}{7}$ for π.

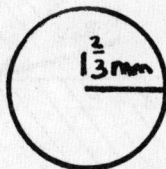

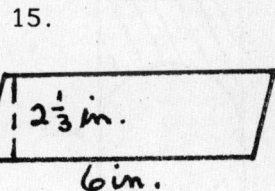

$1\frac{3}{3}$ mm

15.

$2\frac{1}{3}$ in.

6 in.

Find the area.   Remember to include the units.   Draw a picture, if that helps.

16. A square 35 ft on a side.

17. A parallelogram with height of 14 in. and base of 23 in.

18. A trapezoid with height of 5 m and parallel bases of 3 m and 10 m.

19. A circle with radius of 6.9 yd. Use 3.14 for π.   Round to the nearest tenth.

20. A rectangle with length of 8.8 m and width of 4.2 m.

21. A triangle with height of 9 km and base of 5.2 km.

22. A trapezoid with bases of 22.3 cm and 35.8 cm and a height of 16.5 cm.   Round to the nearest hundredth.

23. A rectangle $\frac{3}{4}$ mi by $\frac{7}{8}$ mi.

24. A triangle with a base of $\frac{3}{8}$ yd and height of $2\frac{1}{2}$ yd.

25. A circle of radius 1.7 km.   Use 3.14 for π.   Round to the nearest tenth.

26. A parallelogram with base of $9\frac{1}{2}$ ft and height of $8\frac{1}{4}$ ft.

27. A triangle with base of 15.3 cm and height of 9.2 cm.

Find the area of the shaded regions.   Use 3.14 for π.

28.

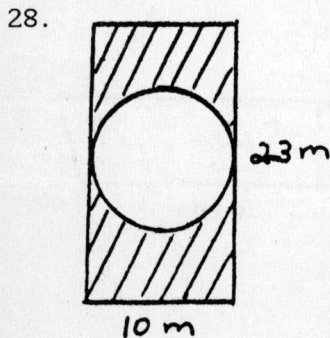

29.

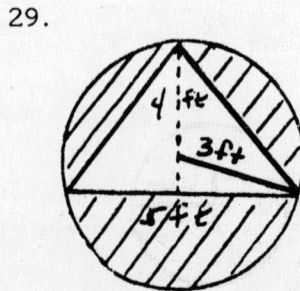

30.

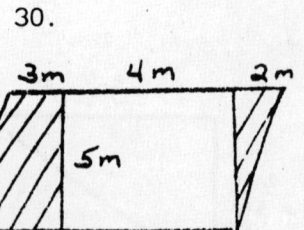

EXTRA PRACTICE 18
Converting Units of Measure
Use after Sections 10.1, 10.3, and 10.5   Name _____

Complete. Recall that the metric prefixes mean the same with liters and grams as they do with meters.

1. 35 L = _____ mL

2. 6.25 L = _____ mL

3. 420 mL = _____ L

4. 56 mL = _____ cm^3

5. 56 mL = _____ L

6. 7820 mL = _____ L

7. 145 cm^3 = ____ mL

8. 145 mL = _____ L

9. 145 cm^3 = _____ L

10. 0.014 L = ____ mL

11. 2500 L = _____ mL

12. 5000 mL = ____ L

13. 5000 mg = ____ g

14. 2500 g = _____ mg

15. 15 g = _____ mg

16. 135 mg = _____ g

17. 3.6 g = _____ mg

18. 5.7 g = _____ cg

19. 942 cg = _____ g

20. 97 cg = _____ mg

21. 5.2 mg = _____ cg

22. 41 kg = _____ g

23. 1600 g = _____ kg

24. 0.55 cg = _____ mg

25. 0.6 g = _____ cg

26. 7545 cg = _____ g

27. 47 g = _____ kg

28. 347 mg = _____ cg

29. 4.9 kg = _____ g

30. 8.4 cg = _____ mg

Complete.

31. 7 lb = _____ oz

32. 56 oz = _____ lb

33. 5000 lb = _____ T

34. 1.25 T = _____ lb

35. 15 lb = _____ oz

36. 160 oz = _____ lb

37. 20$\frac{1}{2}$ lb = _____ oz

38. 1000 lb = _____ T

39. 4 oz = _____ lb

25

Complete.

40. 5 min = _____ sec   41. 3 days = _____ hr   42. 180 min = _____ hr

43. 3 wk = _____ hr   44. 2400 sec = _____ min   45. 9 wk = _____ days

46. 105 days = ____ wk   47. 15 days = _____ hr   48. 84 hr = _____ days

49. $3\frac{1}{2}$ days = ____ wk   50. 3 yr = _____ days   51. 1 yr = _____ wk
(approximately)

52. 1 yr = _____ hr   53. 1 yr = _____ min   54. $\frac{1}{2}$ yr = _____ days

Complete.

55. 3 yd^2 = _____ ft^2 56. 54 ft^2 = _____ yd^2 57. $1\frac{1}{2}$ yd^2 = _____ ft^2

58. 1 ft^2 = _____ yd^2 59. 3 ft^2 = _____ in^2 60. 576 in^2 = _____ ft^2

61. 2 mi^2 = _____ acres   62. 3520 acres = ___ mi^2 63. 3 cm^2 = _____ mm^2

64. 560 mm^2 = _____ cm^2 65. 10 m^2 = _____ cm^2 66. 0.5 m^2 = _____ cm^2

67. 16,000 cm^2 = __ m^2 68. 500,000 m^2 = ___ km^2 69. 6.2 km^2 = _____ m^2

70. 3.5 m^2 = _____ cm^2 71. 5000 m^2 = _____ km^2 72. 425 mm^2 = _____ cm^2

Name _____

EXAMPLES: Add. $4 + 7 = 11$ $4 + (-7) = -3$

$-4 + 7 = 3$ $-4 + (-7) = -11$

Add.

1. $-5 + (-6) =$ _____

2. $-8 + 3 =$ _____

3. $9 + (-4) =$ _____

4. $-5 + 5 =$ _____

5. $-14 + (-13) =$ _____

6. $-12 + 9 =$ _____

7. $-3 + 10 =$ _____

8. $4 + (-5) =$ _____

9. $14 + (-8) =$ _____

10. $-6 + (-15) =$ _____

11. $-7 + 5 =$ _____

12. $2 + (-8) =$ _____

13. $-1 + (-4) =$ _____

14. $-7 + (-7) =$ _____

15. $16 + (-12) =$ _____

16. $-11 + 5 =$ _____

17. $-9 + 16 =$ _____

18. $20 + (-8) =$ _____

19. $18 + (-9) =$ _____

20. $-6 + 7 =$ _____

21. $-4 + (-8) =$ _____

22. $10 + (-6) =$ _____

23. $-9 + 13 =$ _____

24. $-3 + (-8) =$ _____

25. $-16 + 16 =$ _____

26. $-5 + 1 =$ _____

27. $8 + (-15) =$ _____

28. $-4 + (-18) =$ _____

29. $-7 + (-5) =$ _____

30. $-6 + (-6) =$ _____

31. $-2 + 5 =$ _____

32. $1 + (-1) =$ _____

33. $20 + (-14) =$ _____

34. $-18 + 11 =$ _____

35. $-4 + (-3) =$ _____

36. $2 + (-11) =$ _____

37. $-5 + 7 =$ _____

38. $-6 + (-9) =$ _____

39. $12 + (-4) =$ _____

40. $-14 + 5 =$ _____

Name _____

EXAMPLES: Subtract. $3 - 5 = 3 + (-5) = -2$ $3 - (-5) = 3 + 5 = 8$

$-3 - 5 = -3 + (-5) = -8$ $-3 - (-5) = -3 + 5 = 2$

Subtract.

1. $-5 - 9 =$ _____

2. $-6 - (-6) =$ _____

3. $12 - (-2) =$ _____

4. $5 - 6 =$ _____

5. $-4 - 6 =$ _____

6. $-7 - (-10) =$ _____

7. $2 - 8 =$ _____

8. $-9 - 1 =$ _____

9. $3 - (-4) =$ _____

10. $1 - (-14) =$ _____

11. $-5 - (-7) =$ _____

12. $-6 - 7 =$ _____

13. $8 - (-3) =$ _____

14. $-18 - (-9) =$ _____

15. $9 - 16 =$ _____

16. $20 - 8 =$ _____

17. $-11 - 3 =$ _____

18. $2 - (-9) =$ _____

19. $-15 - (-3) =$ _____

20. $7 - 8 =$ _____

21. $4 - (-7) =$ _____

22. $-11 - (-9) =$ _____

23. $2 - 10 =$ _____

24. $5 - (-8) =$ _____

25. $-3 - 1 =$ _____

26. $15 - 20 =$ _____

27. $6 - (-3) =$ _____

28. $-8 - (-10) =$ _____

29. $-5 - 12 =$ _____

30. $11 - (-3) =$ _____

31. $18 - 21 =$ _____

32. $-1 - 1 =$ _____

33. $-7 - (-9) =$ _____

34. $4 - 5 =$ _____

35. $12 - (-8) =$ _____

36. $18 - (-4) =$ _____

37. $-4 - 6 =$ _____

38. $-2 - (-3) =$ _____

39. $10 - 13 =$ _____

40. $7 - (-6) =$ _____

EXTRA PRACTICE 21
Addition and Subtraction of Real Numbers
Use after Sections 11.2 and 11.3   Name _____

Compute.

1. $-6 + 10 =$ _____   2. $-3 + 9 =$ _____   3. $3 - 8 =$ _____

4. $7 - 1 =$ _____   5. $-3 - 5 =$ _____   6. $-4 - 2 =$ _____

7. $-10 - 3 =$ _____   8. $-5 + 2 =$ _____   9. $0 - 12 =$ _____

10. $7 - 2 =$ _____   11. $-10 + 5 =$ _____   12. $-5 + 5 =$ _____

13. $-4 - 3 =$ _____   14. $-3 + 6 =$ _____   15. $-5 - 5 =$ _____

16. $-6 + 14 =$ _____   17. $-4 - 16 =$ _____   18. $13 - 13 =$ _____

19. $-3 + 10 =$ _____   20. $4 - 6 =$ _____   21. $6 - 10 =$ _____

22. $-5 - 3 =$ _____   23. $-7 - 2 =$ _____   24. $-5 - 6 =$ _____

25. $-8 - (-2) =$ _____   26. $11 - (-3) =$ _____   27. $5 - (-8) =$ _____

28. $-9 - (-10) =$ _____   29. $7 - (-3) =$ _____   30. $-7 - 2 =$ _____

31. $12 - 15 =$ _____   32. $-3 - 6 =$ _____   33. $-2 - (-9) =$ _____

34. $9 - (-10) =$ _____   35. $-6 - (-6) =$ _____   36. $-8 - 4 =$ _____

37. $20 - 14 =$ _____   38. $-34 + 4 =$ _____   39. $-40 + 8 =$ _____

40. $-51 - 6 =$ _____   41. $-49 + 16 =$ _____   42. $-38 + 52 =$ _____

EXTRA PRACTICE 21 (continued)
Addition and Subtraction of Real Numbers
Use after Sections 11.2 and 11.3

43. 36 - 90 = _____

44. -52 - 12 = _____

45. 88 - 88 = _____

46. -60 + 60 = _____

47. -54 - 20 = _____

48. -47 + 17 = _____

49. 73 - 18 = _____

50. -55 - 14 = _____

51. -38 + 48 = _____

52. 86 + 14 = _____

53. 66 - 22 = _____

54. -35 - 25 = _____

55. -38 + 67 = _____

56. -56 + 48 = _____

57. 50 - 77 = _____

58. 28 + 61 = _____

59. -82 + 54 = _____

60. -72 - 36 = _____

61. -23 + 55 = _____

62. 52 - 73 = _____

63. 94 - 33 = _____

64. 51 + 26 = _____

65. -94 - 48 = _____

66. -113 + 57 = _____

67. 253 - 151 = _____

68. 206 - 364 = _____

69. -105 - 55 = _____

70. -200 + 333 = _____

71. 400 - 634 = _____

72. 500 - 250 = _____

73. -175 - 225 = _____

74. 333 - 500 = _____

75. -546 + 46 = _____

76. -451 - 239 = _____

77. -758 + 758 = _____

78. -632 + 631 = _____

79. 852 - 902 = _____

80. -444 - 222 = _____

81. 111 - 888 = _____

82. -694 + 695 = _____

83. -624 + 251 = _____

84. 341 - 426 = _____

85. -357 - 192 = _____

86. -1000 + 995 = _____

87. -500 - 2500 = _____

88. 6431 - 7111 = _____

EXTRA PRACTICE 22
Multiplication and Division of Real Numbers
Use after Sections 11.4 and 11.5   Name _____

EXAMPLES: Multiply. $6 \cdot 7 = 42$ $5(-3) = -15$

$-4 \cdot 3 = -12$ $(-8)(-2) = 16$

Divide. $\dfrac{15}{3} = 5$ $\dfrac{40}{-10} = -4$

$\dfrac{-56}{7} = -8$ $\dfrac{-25}{-5} = 5$

Compute.

1. $5(3) = $ _____

2. $6(-4) = $ _____

3. $-2(1) = $ _____

4. $-3(7) = $ _____

5. $(-5)(-4) = $ ____

6. $4(-8) = $ _____

7. $-9(-6) = $ _____

8. $-10(3) = $ ____

9. $6(-3) = $ _____

10. $(-4)(-1) = $ ____

11. $(-1)(-1) = $ ____

12. $-7(-3) = $ ____

13. $9(4) = $ _____

14. $3(-10) = $ _____

15. $-6(-5) = $ _____

16. $30(-2) = $ ____

17. $(-8)(-5) = $ ____

18. $9(6) = $ _____

19. $-3(0) = $ _____

20. $-10(10) = $ ____

21. $\dfrac{20}{5} = $ _____

22. $\dfrac{-15}{3} = $ _____

23. $\dfrac{16}{-4} = $ _____

24. $\dfrac{-12}{6} = $ _____

25. $\dfrac{7}{-1} = $ _____

26. $\dfrac{-8}{-2} = $ _____

27. $\dfrac{-10}{-5} = $ _____

28. $\dfrac{14}{-7} = $ _____

29. $\dfrac{18}{6} = $ _____

30. $\dfrac{-3}{-1} = $ _____

31. $\dfrac{25}{-5} = $ _____

32. $\dfrac{0}{-5} = $ _____

33. $\dfrac{-6}{-3} = $ _____

34. $\dfrac{27}{-9} = $ _____

35. $\dfrac{36}{6} = $ _____

36. $\dfrac{-54}{9} = $ _____

37. $\dfrac{-81}{-9} = $ _____

38. $\dfrac{72}{-8} = $ _____

39. $\dfrac{-35}{-7} = $ _____

40. $\dfrac{30}{3} = $ _____

41. $-10(16) = $ _____

42. $12(-11) = $ _____

43. $-15(5) = $ _____

44. $(-7)(-20) = $ __

45. $15(13) = $ _____

46. $-14(-21) = $ _____

47. $22(-8) = $ _____

48. $-56(-23) = $ ___

EXTRA PRACTICE 22 (continued)
Multiplication and Division of Real Numbers
Use after Sections 11.4 and 11.5

49. $(-50)(130) = $ _____

50. $60(-160) = $ _____

51. $-1(540) = $ _____

52. $0(-326) = $ _____

53. $(-1)(-256) = $ _____

54. $(10)(-17) = $ _____

55. $-100(-30) = $ _____

56. $-15(15) = $ _____

57. $20(14) = $ _____

58. $-36(1) = $ _____

59. $(-30)(-30) = $ _____

60. $-369(0) = $ _____

61. $50(25) = $ _____

62. $-6(35) = $ _____

63. $(-4)(-25) = $ _____

64. $(20)(-24) = $ _____

65. $-2(-360) = $ _____

66. $-11(26) = $ _____

67. $54(-22) = $ _____

68. $\dfrac{-50}{-25} = $ _____

69. $\dfrac{-75}{3} = $ _____

70. $\dfrac{120}{-40} = $ _____

71. $\dfrac{-165}{-1} = $ _____

72. $\dfrac{435}{-435} = $ _____

73. $\dfrac{-520}{130} = $ _____

74. $\dfrac{408}{-34} = $ _____

75. $\dfrac{-1111}{-11} = $ _____

76. $\dfrac{-500}{10} = $ _____

77. $\dfrac{-1600}{-100} = $ _____

78. $\dfrac{0}{-371} = $ _____

79. $\dfrac{720}{-36} = $ _____

80. $\dfrac{480}{120} = $ _____

81. $\dfrac{-256}{-16} = $ _____

82. $\dfrac{360}{-18} = $ _____

83. $\dfrac{-399}{19} = $ _____

84. $\dfrac{-3478}{-3478} = $ _____

85. $\dfrac{-444}{111} = $ _____

86. $\dfrac{2300}{-100} = $ _____

87. $\dfrac{569}{-1} = $ _____

88. $\dfrac{-240}{15} = $ _____

89. $\dfrac{4100}{-10} = $ _____

90. $\dfrac{-800}{-25} = $ _____

91. $\dfrac{-666}{222} = $ _____

92. $\dfrac{500}{-2} = $ _____

Addition, Subtraction, Multiplication, and Division of Real Numbers
Use after Sections 11.2 - 11.5   Name _____

EXAMPLES: Add. $-\frac{3}{4} + \frac{2}{3} = -\frac{9}{12} + \frac{8}{12} = -\frac{1}{12}$

$$4.3 + (-6.2) = -1.9$$

Add.

1. $-\frac{5}{6} + \left(-\frac{1}{3}\right) =$ _____

2. $\frac{5}{8} + \left(-\frac{2}{5}\right) =$ _____

3. $-\frac{7}{12} + \frac{1}{8} =$ _____

4. $-\frac{1}{4} + \left(-\frac{1}{6}\right) =$ _____

5. $\frac{5}{9} + \left(-\frac{3}{8}\right) =$ _____

6. $-\frac{2}{3} + \frac{9}{10} =$ _____

7. $-8.2 + 2.3 =$ _____

8. $-6.4 + 11.3 =$ _____

9. $-5.6 + (-4.3) =$ _____

10. $12.7 + (-15.4) =$ _____

11. $-4.9 + 8.7 =$ _____

12. $-2.8 + (-9.9) =$ _____

EXAMPLES: Subtract. $-\frac{3}{4} - \frac{2}{3} = -\frac{9}{12} + \left(-\frac{8}{12}\right) = -\frac{17}{12}$

$$5.8 - (-2.3) = 5.8 + 2.3 = 8.1$$

13. $\frac{7}{10} - \frac{3}{2} =$ _____

14. $-\frac{1}{3} - \frac{1}{6} =$ _____

15. $-\frac{5}{8} - \left(-\frac{2}{3}\right) =$ _____

16. $\frac{5}{12} - \left(-\frac{2}{5}\right) =$ _____

17. $-\frac{3}{7} - \frac{3}{14} =$ _____

18. $-\frac{4}{9} - \left(-\frac{5}{6}\right) =$ _____

19. $6.2 - (-12.9) =$ _____

20. $-3.5 - 15.4 =$ _____

21. $4.9 - 8.5 =$ _____

22. $-10.4 - (-6.6) =$ _____

23. $-6.7 - 2.4 =$ _____

24. $5.8 - (-2.8) =$ _____

EXAMPLES: Multiply. $-\frac{3}{4} \cdot \left(\frac{2}{3}\right) = -\frac{1}{2}$

$(-4.2) \cdot (-3.8) = 15.96$

25. $\frac{3}{5} \cdot \left(-\frac{10}{9}\right) =$ _____

26. $-\frac{7}{8} \cdot \frac{12}{5} =$ _____

27. $-\frac{5}{8} \cdot \left(-\frac{16}{25}\right) =$ _____

28. $\frac{3}{10} \cdot \left(-\frac{20}{21}\right) =$ _____

29. $-\frac{12}{5} \cdot \frac{20}{3} =$ _____

30. $-\frac{5}{9} \cdot \left(-\frac{18}{35}\right) =$ _____

31. $(0.2) \cdot (-0.3) =$ _____

32. $(-2.3) \cdot (3.1) =$ _____

33. $(-5.6) \cdot (-4.1) =$ _____

34. $(1.3) \cdot (-6.2) =$ _____

35. $(-1.2) \cdot (2.4) =$ _____

36. $(-3.1) \cdot (-8.8) =$ _____

EXAMPLES: Divide. $-\frac{3}{4} \div -\frac{2}{3} = -\frac{3}{4} \cdot \left(-\frac{3}{2}\right) = \frac{9}{8}$

$-6.9 \div 2.3 = -3$

37. $\frac{7}{8} \div \left(-\frac{3}{16}\right) =$ _____

38. $-\frac{7}{9} \div \frac{28}{27} =$ _____

39. $-\frac{5}{6} \div \left(-\frac{5}{24}\right) =$ _____

40. $\frac{8}{9} \div \left(-\frac{16}{15}\right) =$ _____

41. $-\frac{1}{3} \div \frac{1}{3} =$ _____

42. $-\frac{3}{7} \div \left(-\frac{9}{14}\right) =$ _____

43. $14.7 \div (-2.1) =$ _____

44. $-8.8 \div 2.2 =$ _____

45. $-16.5 \div (-3.3) =$ _____

46. $12.1 \div (-1.1) =$ _____

47. $-2.56 \div 1.6 =$ _____

48. $-18.9 \div 6.3 =$ _____

Simplifying Expressions and Order of Operations
Use after Section 11.5   Name _____

EXAMPLE: Simplify. $\dfrac{2^2 - 3 \cdot 4 + 7}{5 - 2^2 \cdot 3 + 6} = \dfrac{4 - 3 \cdot 4 + 7}{5 - 4 \cdot 3 + 6}$

$$= \dfrac{4 - 12 + 7}{5 - 12 + 6}$$

$$= \dfrac{-1}{-1}$$

$$= 1$$

1. $2^3 - 3^2 =$ _____

2. $2 \cdot 3 - 4 \cdot 2 + 7 =$ _____

3. $5(-1) + 6(-2) =$ _____

4. $(-2)(3) - (-1)(7) - (-2) =$ ___

5. $3 + 2^2 - 16 \cdot 3^2 =$ _____

6. $6 + (3 - 4) - 2 =$ _____

7. $6 + 3 - (4 - 2) =$ _____

8. $3^2 - 8 \cdot 2 + 7^2 - 35 =$ _____

9. $-4(2^3) - 6 =$ _____

10. $(8 - 2)^2 =$ _____

11. $(4 - 6)^2 =$ _____

12. $4 - 6^2 =$ _____

13. $[32 \div (-4)] \div 2 =$ _____

14. $32 \div [(-4) \div 2] =$ _____

15. $\dfrac{4 - 3^2}{8^2 + 2} =$ _____

16. $\dfrac{7^2 - 8^2 + 1^3}{2^3 + 3^2 - 2^3} =$ _____

17. $\dfrac{2(8 + 3) - 4(7 + 2)}{5(6 - 1) - 3(8 - 6)} =$ __

18. $\dfrac{3.2(2 - 4) + 6}{4.9 - 3(6 + 1)} =$ _____

19. $\dfrac{8 - 4^2 + 3 \cdot 5}{4 \cdot 2 - 3^2 + 9} =$ _____

20. $\dfrac{2 \cdot 3 - 4 \cdot 5 + 6}{-20 \div (-5) \div 8} =$ _____

Multiply using the distributive law.

EXAMPLE: $8(6x - 4) = 8 \cdot 6x - 8 \cdot 4 = 48x - 32$

Multiply.

1. $7(n - 4) =$ _____   2. $5(x + 6) =$ _____

3. $-2(x - 7) =$ _____   4. $-9(y + 10) =$ _____

5. $-4(a + 3b) =$ _____   6. $10(2x - 3y) =$ _____

7. $-5(x + 2y - 6) =$ _____   8. $8(5x + 4y - 12) =$ _____

9. $9(2a - b + 3) =$ _____   10. $-7(-3p - 15q + 14) =$ _____

11. $3(-6r + 15t - 21) =$ _____   12. $20(-6a - 10b + 9) =$ _____

Factoring is the reverse of multiplying. To factor, we can use the distributive laws in reverse.

EXAMPLE: $3x + 27y - 6 = 3 \cdot x + 3 \cdot 9y - 3 \cdot 2 = 3(x + 9y - 2)$

Factor.

13. $6x - 6 =$ _____   14. $8x + 24 =$ _____

15. $4x - 28 =$ _____   16. $5y - 30 =$ _____

17. $7x + 7 =$ _____   18. $9x - 63 =$ _____

19. $48 - 8x =$ _____   20. $55 - 11x =$ _____

21. $6a + 9 =$ _____   22. $14x - 49 =$ _____

23. $10y + 15 =$ _____   24. $18a - 30 =$ _____

25. $50x - 70 =$ _____   26. $32x + 24 =$ _____

27. $6x + 30 - 36a =$ _____   28. $15x + 45y - 15 =$ _____

29. $bx - 3by + 6b =$ _____   30. $ax - 5a - ay =$ _____

31. $6x + 9y - 24 =$ _____   32. $25x - 15y + 75 =$ _____

Solving Equations Using the Addition and Multiplication Principles
Use after Sections 12.2 - 12.4   Name _____

Solve.

1. $x + 37 = 98$ _____

2. $y - 53 = 141$ _____

3. $59 + a = -123$ _____

4. $-72 + t = -40$ _____

5. $-55 = x + 32$ _____

6. $a + \dfrac{5}{6} = -\dfrac{1}{2}$ _____

7. $\dfrac{3}{4} + x = \dfrac{7}{8}$ _____

8. $y - 3\dfrac{1}{2} = -2\dfrac{2}{3}$ _____

9. $48x = -192$ _____

10. $-25a = -200$ _____

11. $-15y = 96$ _____

12. $-\dfrac{1}{3}x = 48$ _____

13. $\dfrac{3}{2}r = -\dfrac{4}{5}$ _____

14. $x - 56 = -42$ _____

15. $15 - y = 33$ _____

16. $51 - x = -133$ _____

17. $-31t = -93$ _____

18. $-53 + a = 65$ _____

19. $-\dfrac{5}{3}b = -\dfrac{1}{6}$ _____

20. $58x = -145$ _____

21. $-89 = -27 - a$ _____

22. $\dfrac{x}{4} = -45$ _____

23. $\dfrac{r}{-3} = \dfrac{1}{3}$ _____

24. $\dfrac{11}{2}y = -3\dfrac{2}{3}$ _____

25. $t + \dfrac{5}{8} = -\dfrac{3}{4}$ _____

26. $\dfrac{b}{-5} = 11$ _____

27. $-\dfrac{7}{8}t = -\dfrac{7}{8}$ _____

28. $3x + 5x = 48$ _____

29. $18x - 12x = -96$ _____

30. $3y - 13y = 50$ _____

31. $9t - 16t = -49$ _____

32. $5a - 4 = 26$ _____

33. $8r + 16 = -48$ _____

34. $-10x - 41 = 69$ _____

35. $11b = 45 - 4b$ _____

36. $9z + \frac{1}{2}z = 38$ _____

37. $x + 58 = 135$ _____

38. $62y = -558$ _____

39. $3a + 4a - 3 = 11$ _____

40. $6x + 5 - 2x = -19$ _____

41. $9r + 3r - 5 = 25$ _____

42. $3x + 2 = 2x - 6$ _____

43. $5z - 4 = 4z - 3$ _____

44. $4y + 2y - 7 = 3y + 11$ _____

45. $3t - 5 = 7t + t - 15$ _____

46. $6x + 5x - 4 = 2x - 8$ _____

47. $\frac{1}{2}x + \frac{1}{3}x = \frac{1}{6}x - 5$ _____

48. $\frac{2}{3}y - \frac{5}{4}y + 8 = -\frac{11}{12}y - 4$ _____

49. $\frac{z}{-5} = -15$ _____

50. $\frac{t}{2} = -33$ _____

51. $\frac{h}{13} = 0$ _____

52. $-2y + 7 = 7$ _____

53. $5x - 4 = 4x - 4$ _____

54. $-3462a = 0$ _____

Example: Three plus six times a number is 7 more than four times the number. What is the number?

Three plus six times a number is 7 more than four times the number.

$$3 + 6 \cdot x = 7 + 4 \cdot x$$

Solve: $3 + 6x = 7 + 4x$

$3 + 2x = 7$

$2x = 4$

$x = 2$

Solve.

1. When 6 is added to three times a number, the result is 30. Find the number. _____

2. When you double a number and then add 20, you get $\frac{4}{3}$ of the original number. Find the number. _____

3. The perimeter of a rectangle is 52 cm. The length is 8 cm greater than the width. Find the width and length. _____

4. The perimeter of a rectangle is 78 m. The width is 7 m less than the length. Find the length and width. _____

5. The sum of three consecutive even integers is 150. Find the integers. _____

6. The sum of three consecutive odd integers is 261. Find the integers.

   _____

7. A 20 ft board is cut into three pieces. The second piece is three times as long as the first. The third piece is twice as long as the second. Find the lengths of the pieces. _____

8. A 450 m fence is divided into three sections. The second section is twice as long as the first. The third section is three times as long as the second. Find the lengths of the sections. _____

9. The second angle of a triangle is three times as large as the first. The third angle is 20° larger than the sum of the first two. Find the measures of the angles. _____

10. The second angle of a triangle is twice as large as the first. The third angle is 50° less than the second. Find the measures of the angles.

    _____

11. The cost of renting a car is $18 per day plus 16¢ per mile. Find the cost of renting a car for a three-day trip of 1000 miles. _____

12. Thirteen less than twice a number is seventeen more than half the number. What is the number? _____

ANSWER KEYS FOR EXTRA PRACTICE SHEETS

Extra Practice 1

<u>1</u>. 39   <u>2</u>. 47   <u>3</u>. 55   <u>4</u>. 74   <u>5</u>. 102   <u>6</u>. 495   <u>7</u>. 1189   <u>8</u>. 964

<u>9</u>. 1472   <u>10</u>. 1616   <u>11</u>. 11,154   <u>12</u>. 4294   <u>13</u>. 103,859   <u>14</u>. 809,402

<u>15</u>. 1,217,745   <u>16</u>. 243   <u>17</u>. 511   <u>18</u>. 239   <u>19</u>. 464   <u>20</u>. 2279

<u>21</u>. 708   <u>22</u>. 4547   <u>23</u>. 86   <u>24</u>. 13,904   <u>25</u>. 19,819   <u>26</u>. 15,982

<u>27</u>. cannot be done using only positive numbers   <u>28</u>. 64,609   <u>29</u>. 50,295

<u>30</u>. 360,506

Extra Practice 2

<u>1</u>. 1674   <u>2</u>. 1155   <u>3</u>. 3381   <u>4</u>. 9483   <u>5</u>. 301,416   <u>6</u>. 634,073

<u>7</u>. 342,925   <u>8</u>. 513,486   <u>9</u>. 23,769,860   <u>10</u>. 38,033,441   <u>11</u>. 6,331,182

<u>12</u>. 78,939,280   <u>13</u>. 153,062   <u>14</u>. 5,087,278   <u>15</u>. 9728   <u>16</u>. 156,924

<u>17</u>. 861 R 2   <u>18</u>. 2 R 13   <u>19</u>. 2 R 9   <u>20</u>. 4 R 18   <u>21</u>. 11 R 29

<u>22</u>. 22 R 17   <u>23</u>. 60 R 46   <u>24</u>. 4 R 45   <u>25</u>. 3 R 45   <u>26</u>. 9 R 305

<u>27</u>. 7 R 128   <u>28</u>. 9 R 812   <u>29</u>. 4 R 748   <u>30</u>. 1 R 316   <u>31</u>. 3 R 2485

<u>32</u>. 11 R 245

Extra Practice 3

<u>1</u>. 11   <u>2</u>. $59,500   <u>3</u>. 483   <u>4</u>. 1030   <u>5</u>. 157,399 square miles

<u>6</u>. 1050   <u>7</u>. 71,236   <u>8</u>. 7 decibels   <u>9</u>. 7 bills with no change

<u>10</u>. 21 bowls with 1 ounce left   <u>11</u>. 648,038   <u>12</u>. 7   <u>13</u>. 6 pounds

<u>14</u>. 2208 years   <u>15</u>. 8   <u>16</u>. 48,681,000 miles   <u>17</u>. $1420   <u>18</u>. 45

<u>19</u>. 20 days

Extra Practice 4

<u>1</u>. 6   <u>2</u>. 38   <u>3</u>. 14   <u>4</u>. 3   <u>5</u>. 15   <u>6</u>. 63   <u>7</u>. 50   <u>8</u>. 20   <u>9</u>. 53

<u>10</u>. 2   <u>11</u>. 0   <u>12</u>. 5   <u>13</u>. 150   <u>14</u>. 0   <u>15</u>. 535   <u>16</u>. 100   <u>17</u>. 4

<u>18</u>. 48   <u>19</u>. 5   <u>20</u>. 2   <u>21</u>. 9   <u>22</u>. 9   <u>23</u>. 4   <u>24</u>. 140

Extra Practice 5

1. Factors: 1, 2, 3, 6; Multiples: 6, 12, 18, 24   2. Factors: 1, 2, 3, 4, 6, 12; Multiples: 12, 24, 36, 48   3. Factors: 1, 2, 4, 5, 10, 20; Multiples: 20, 40, 60, 80   4. Factors: 1, 3, 5, 9, 15, 45; Multiples: 45, 90, 135, 180   5. Factors: 1, 2, 5, 10, 25, 50; Multiples: 50, 100, 150, 200   6. Factors: 1, 3, 7, 21; Multiples: 21, 42, 63, 84   7. Factors: 1, 2, 3, 6, 9, 18; Multiples: 18, 36, 54, 72   8. Factors: 1, 2, 3, 5, 6, 10, 15, 30; Multiples: 30, 60, 90, 120   9. Factors: 1, 3, 9, 27; Multiples: 27, 54, 81, 108   10. Factors: 1, 5, 7, 35; Multiples: 35, 70, 105, 140   11. Factors: 1, 2, 4, 7, 8, 14, 28, 56; Multiples: 56, 112, 168, 224   12. Factors: 1, 3, 13, 39; Multiples: 39, 78, 117, 156   13. Factors: 1, 2, 4, 5, 8, 10, 16, 20, 40, 80; Multiples: 80, 160, 240, 320   14. Factors: 1, 37; Multiples: 37, 74, 111, 148   15. Factors: 1, 5, 11, 55; Multiples: 55, 110, 165, 220   16. Factors: 1, 53; Multiples: 53, 106, 159, 212   17. Factors: 1, 2, 4, 5, 10, 20, 25, 50, 100; Multiples: 100, 200, 300, 400   18. Factors: 1, 5, 25, 125; Multiples: 125, 250, 375, 500   19. Factors: 1, 2, 4, 8, 16, 32, 64, 128; Multiples: 128, 256, 384, 512   20. Factors: 1, 3, 37, 111; Multiples: 111, 222, 333, 444

Extra Practice 6

1. 2, 3, 6, 7, 9   2. 3, 5   3. 2, 5, 10   4. 5   5. 2, 4   6. 7   7. 2, 4, 5, 7, 8, 10   8. 2, 3, 4, 5, 6, 9, 10   9. 2, 3, 4, 6, 9   10. 2, 3, 4, 6, 7, 8   11. 3, 5, 9   12. 3, 9   13. 2, 3, 4, 5, 6, 8, 10   14. 2, 3, 4, 6   15. 2, 3, 6   16. 2, 4, 8   17. 2, 3, 4, 5, 6, 8, 10   18. 2, 3, 5, 6, 9, 10   19. none   20. 2

Extra Practice 7

1. $2 \cdot 3 \cdot 3 \cdot 3 \cdot 7$ 2. $3 \cdot 5 \cdot 11 \cdot 19$ 3. $2 \cdot 5 \cdot 5 \cdot 11$ 4. $5 \cdot 5 \cdot 5 \cdot 5$ 5. $2 \cdot 2 \cdot 11 \cdot 13$ 6. $7 \cdot 11 \cdot 13$ 7. $2 \cdot 2 \cdot 2 \cdot 5 \cdot 7$ 8. $2 \cdot 2 \cdot 3 \cdot 3 \cdot 3 \cdot 5$ 9. $2 \cdot 2 \cdot 3 \cdot 3 \cdot 17$ 10. $2 \cdot 2 \cdot 2 \cdot 3 \cdot 7 \cdot 11$

<u>11</u>. $3 \cdot 3 \cdot 3 \cdot 3 \cdot 5 \cdot 5$ <u>12</u>. $3 \cdot 3 \cdot 3 \cdot 19$ <u>13</u>. $2 \cdot 2 \cdot 2 \cdot 3 \cdot 5 \cdot 11$

<u>14</u>. $2 \cdot 2 \cdot 3 \cdot 31$ <u>15</u>. $2 \cdot 3 \cdot 23$ <u>16</u>. $2 \cdot 2 \cdot 2 \cdot 2 \cdot 2 \cdot 2 \cdot 2$

<u>17</u>. $2 \cdot 2 \cdot 2 \cdot 3 \cdot 5 \cdot 5 \cdot 5$ <u>18</u>. $2 \cdot 3 \cdot 3 \cdot 5 \cdot 5 \cdot 19$ <u>19</u>. prime

<u>20</u>. $2 \cdot 73$

Extra Practice 8

<u>1</u>. $\frac{21}{32}$ <u>2</u>. 2 <u>3</u>. $\frac{7}{5}$ <u>4</u>. $\frac{5}{4}$ <u>5</u>. $\frac{20}{3}$ <u>6</u>. $\frac{20}{7}$ <u>7</u>. $\frac{11}{6}$ <u>8</u>. $\frac{3}{28}$ <u>9</u>. $\frac{16}{33}$

<u>10</u>. $\frac{3}{2}$ <u>11</u>. $\frac{1}{15}$ <u>12</u>. $\frac{2}{3}$ <u>13</u>. 13 <u>14</u>. $\frac{1}{13}$ <u>15</u>. $\frac{93}{70}$ <u>16</u>. $\frac{3}{16}$ <u>17</u>. $\frac{1}{6}$

<u>18</u>. $\frac{19}{30}$ <u>19</u>. $\frac{2}{3}$ <u>20</u>. $\frac{31}{7}$ <u>21</u>. $\frac{119}{100}$ <u>22</u>. $\frac{3}{20}$ <u>23</u>. 20 <u>24</u>. $\frac{19}{24}$ <u>25</u>. $\frac{1}{3}$

<u>26</u>. $\frac{1}{7}$ <u>27</u>. 7 <u>28</u>. $\frac{1}{10}$ <u>29</u>. $\frac{1}{48}$ <u>30</u>. $\frac{52}{5}$

Extra Practice 9

<u>1</u>. $9\frac{5}{7}$ <u>2</u>. $3\frac{4}{15}$ <u>3</u>. $8\frac{1}{2}$ <u>4</u>. $1\frac{3}{4}$ <u>5</u>. $3\frac{2}{15}$ <u>6</u>. $9\frac{1}{5}$ <u>7</u>. $13\frac{5}{6}$

<u>8</u>. 8 <u>9</u>. $4\frac{5}{7}$ <u>10</u>. $10\frac{1}{5}$ <u>11</u>. $18\frac{11}{16}$ <u>12</u>. 12 <u>13</u>. 6 <u>14</u>. $6\frac{1}{12}$

<u>15</u>. $5\frac{7}{12}$ <u>16</u>. $4\frac{3}{16}$ <u>17</u>. $17\frac{3}{8}$ <u>18</u>. 35 <u>19</u>. $5\frac{5}{44}$ <u>20</u>. 1 <u>21</u>. 0

<u>22</u>. $41\frac{3}{8}$ <u>23</u>. $36\frac{13}{24}$ <u>24</u>. $48\frac{4}{7}$ <u>25</u>. $61\frac{1}{5}$ <u>26</u>. $13\frac{13}{21}$ <u>27</u>. 0 <u>28</u>. $\frac{47}{60}$

<u>29</u>. $26\frac{8}{9}$ <u>30</u>. $128\frac{1}{4}$

Extra Practice 10

<u>1</u>. $\frac{7}{12}$ c pancake mix, $\frac{5}{12}$ c water, $\frac{1}{3}$ egg, $\frac{1}{6}$ c bran cereal, and $\frac{2}{3}$ T oil.

For $\frac{7}{12}$ c use a little more than $\frac{1}{2}$ c and for $\frac{5}{12}$ c, use a little less than $\frac{1}{2}$ c.

One way to measure the egg would be to beat it and measure.

<u>2</u>. tamale pie, $\frac{1}{5}$ <u>3</u>. $\frac{1}{4}$, $12\frac{1}{2}$ lb   <u>4</u>. 141,667   <u>5</u>. \$6230   <u>6</u>. $\$27\frac{3}{4}$

Extra Practice 10 (continued)

7. $\frac{7}{24}$ yd   8. $\frac{1}{12}$ ft   9. 17 mi   10. 40 g   11. $\frac{5}{8}$ 12. $\frac{3}{8}$ lb, more

13. 9   14. $12,800   15. $285   16. 3 in.   17. 20   18. $\frac{9}{16}$ in., $\frac{1}{16}$ in.

19. $4\frac{1}{2}$ acres   20. less, $\frac{8}{9}$ yd

Extra Practice 11

1. $902.17   2. $526.09   3. 7.9   4. 233.4   5. 15.3   6. $17.50

7. 40   8. 45.5 in.   9. Yes, 0.2 m left   10. 1.31 g   11. 15

12. Belinda, 7¢   13. $27   14. 0.6   15. 43   16. $171.50   17. 7.5

18. 880   19. 25,000   20. 0.012

Extra Practice 12

1. 18 days   2. 510 miles   3. 38   4. 8   5. 62.5   6. 11 in.

7. $7\frac{1}{2}$ cups   8. 8.5   9. 156   10. $\frac{3}{4}$ 11. 24   12. 30.48   13. 88

14. 12   15. $34   16. $x = 25.5$ ft   17. $x = 4$ m

Extra Practice 13

1. $2775   2. 12 ounces   3. 40   4. 7%   5. 1300   6. $4.73   7. 30%

8. 5.25% or $5\frac{1}{4}$%   9. 200 mg   10. 26%   11. $15.60   12. 120 million

13. $\frac{3}{4}$ million   14. $811.20   15. $28, $3158   16. $132.3 million

17. 600%   18. $2520

Extra Practice 14

1. Average: 7; median: 8; mode: 8   2. Average: 40; median: 39.5;

mode: 25, 32, 47, 56   3. Average: 3; median: 2.6; mode: 2.6

4. Average: 371.8; median: 371; mode: 371   5. Average: 27; median: 28;

Extra Practice 14 (continued)

mode: 29   6. Average: $40.07; median: $42.305; mode: $48.36

7. Average: 10.5; median: 10.5; mode: 3, 6, 9, 12, 15, 18

8. Average: 75; median: 79; mode: 79   9. Average: 88; median: 90; mode: 85

10. Average: 79°; median: 79°; mode: 75°   11. 86   12. 46   13. 96

14. 2.80   15. 2.50

Extra Practice 15

1. 180   2. 444   3. 129   4. 378   5. 42   6. $3\frac{1}{4}$ or 3.25   7. 21,120   8. 216

9. 2   10. $\frac{3}{4}$ or 0.75   11. $\frac{2}{3}$ or 0.$\overline{6}$ 12. 8800   13. $\frac{1}{4}$ or 0.25   14. $2\frac{1}{2}$ or 2.5

15. $4\frac{2}{3}$ or 4.$\overline{6}$ 16. 52,800   17. 528   18. 2640   19. $2\frac{1}{4}$ or 2.25   20. 3

21. $\frac{1}{3}$ or 0.$\overline{3}$ 22. 3000   23. 100   24. $\frac{1}{6}$ or 0.1$\overline{6}$ 25. $\frac{1}{3}$ or 0.$\overline{3}$ 26. 2

27. $\frac{1}{2}$ or 0.5   28. 108   29. $3\frac{1}{2}$ or 3.5   30. $12\frac{1}{3}$ or 12.$\overline{3}$

Extra Practice 16

1. 5300   2. 600   3. 3000   4. 89   5. 56.7   6. 0.03   7. 15.2   8. 0.675

9. 3800   10. 0.1   11. 1525   12. 98,000   13. 19.427   14. 0.041   15. 6.2

16. 0.8   17. 45,900   18. 0.0459   19. 775   20. 6   21. 0.7   22. 0.5

23. 0.01   24. 0.52   25. 4.2   26. 0.005   27. 250   28. 528   29. 16.417

30. 94.23

Extra Practice 17

1. 312 ft^2 2. 76 cm^2 3. 803.84 in^2 4. $\frac{11}{14}$ mi^2 5. 31.36 m^2

6. 150.5 mm^2 7. 40.56 ft^2 8. 96 in^2 9. $\frac{1}{5}$ km^2 10. 36.3 m^2

11. $\frac{7}{8}$ cm^2 12. $4\frac{21}{25}$ mi^2 13. $4\frac{1}{3}$ ft^2 14. $8\frac{46}{63}$ mm^2 15. 14 in.^{2}

Extra Practice 17 (continued)

16. 1225 ft^2 17. 322 in^2 18. 32.5 m^2 19. 149.5 yd^2 20. 36.96 m^2

21. 23.4 km^2 22. 479.33 cm^2 23. $\frac{21}{32}$ mi^2 24. $\frac{15}{32}$ yd^2 25. 9.1 km^2

26. 78$\frac{3}{8}$ ft^2 27. 70.38 cm^2 28. 151.5 m^2 29. 18.26 ft^2 30. 25 m^2

Extra Practice 18

1. 35,000   2. 6250   3. 0.42   4. 56   5. 0.056   6. 7.82   7. 145   8. 0.145

9. 0.145   10. 14   11. 2,500,000   12. 5   13. 5   14. 2,500,000   15. 15,000

16. 0.135   17. 3600   18. 570   19. 9.42   20. 970   21. 0.52   22. 41,000

23. 1.6   24. 5.5   25. 60   26. 75.45   27. 0.047   28. 34.7   29. 4900

30. 84   31. 112   32. 3.5 or 3$\frac{1}{2}$ 33. 2.5 or 2$\frac{1}{2}$ 34. 2500   35. 240   36. 10

37. 328   38. 0.5 or $\frac{1}{2}$ 39. 0.25 or $\frac{1}{4}$ 40. 300   41. 72   42. 3   43. 504

44. 40   45. 63   46. 15   47. 360   48. 3.5 or 3$\frac{1}{2}$ 49. 0.5 or $\frac{1}{2}$ 50. 1095$\frac{3}{4}$
or 1095.75   51. 52   52. 8766   53. 525,960   54. 182$\frac{5}{8}$ or 182.625   55. 27

56. 6   57. 13$\frac{1}{2}$ or 13.5   58. $\frac{1}{9}$ 59. 432   60. 4   61. 1280   62. 5.5 or 5$\frac{1}{2}$

63. 300   64. 5.6   65. 100,000   66. 5000   67. 1.6   68. 0.5   69. 6,200,000

70. 35,000   71. 0.005   72. 4.25

Extra Practice 19

1. -11   2. -5   3. 5   4. 0   5. -27   6. -3   7. 7   8. -1   9. 6

10. -21   11. -2   12. -6   13. -5   14. -14   15. 4   16. -6   17. 7

18. 12   19. 9   20. 1   21. -12   22. 4   23. 4   24. -11   25. 0

26. -4   27. -7   28. -22   29. -12   30. -12   31. 3   32. 0   33. 6

34. -7   35. -7   36. -9   37. 2   38. -15   39. 8   40. -9

Extra Practice 20

1. -14 2. 0 3. 14 4. -1 5. -10 6. 3 7. -6 8. -10 9. 7

10. 15 11. 2 12. -13 13. 11 14. -9 15. -7 16. 12 17. -14

18. 11 19. -12 20. -1 21. 11 22. -2 23. -8 24. 13 25. -4

26. -5 27. 9 28. 2 29. -17 30. 14 31. -3 32. -2 33. 2

34. -1 35. 20 36. 22 37. -10 38. 1 39. -3 40. 13

Extra Practice 21

1. 4 2. 6 3. -5 4. 6 5. -8 6. -6 7. -13 8. -3 9. -12

10. 5 11. -5 12. 0 13. -7 14. 3 15. -10 16. 8 17. -20

18. 0 19. 7 20. -2 21. -4 22. -8 23. -9 24. -11 25. -6

26. 14 27. 13 28. 1 29. 10 30. -9 31. -3 32. -9 33. 7

34. 19 35. 0 36. -12 37. 6 38. -30 39. -32 40. -57

41. -33 42. 14 43. -54 44. -64 45. 0 46. 0 47. -74

48. -30 49. 55 50. -69 51. 10 52. 100 53. 44 54. -60

55. 29 56. -8 57. -27 58. 89 59. -28 60. -108 61. 32

62. -21 63. 61 64. 77 65. -142 66. -56 67. 102 68. -158

69. -160 70. 133 71. -234 72. 250 73. -400 74. -167

75. -500 76. -690 77. 0 78. -1 79. -50 80. -666 81. -777

82. 1 83. -373 84. -85 85. -549 86. -5 87. -3000 88. -680

Extra Practice 22

1. 15 2. -24 3. -2 4. -21 5. 20 6. -32 7. 54 8. -30

9. -18 10. 4 11. 1 12. 21 13. 36 14. -30 15. 30 16. -60

17. 40 18. 54 19. 0 20. -100 21. 4 22. -5 23. -4 24. -2

25. -7 26. 4 27. 2 28. -2 29. 3 30. 3 31. -5 32. 0

33. 2 34. -3 35. 6 36. -6 37. 9 38. -9 39. 5 40. 10

41. -160 42. -132 43. -75 44. 140 45. 195 46. 294 47. -176

48. 1288 49. -6500 50. -9600 51. -540 52. 0 53. 256

54. -170 55. 3000 56. -225 57. 280 58. -36 59. 900 60. 0

61. 1250 62. -210 63. 100 64. -480 65. 720 66. -286

67. -1188 68. 2 69. -25 70. -3 71. 165 72. -1 73. -4

74. -12 75. 101 76. -50 77. 16 78. 0 79. -20 80. 4

81. 16 82. -20 83. -21 84. 1 85. -4 86. -23 87. -569

88. -16 89. -410 90. 32 91. -3 92. -250

Extra Practice 23

1. $-\dfrac{7}{6}$ 2. $\dfrac{9}{40}$ 3. $-\dfrac{11}{24}$ 4. $-\dfrac{5}{12}$ 5. $\dfrac{13}{72}$ 6. $\dfrac{7}{30}$ 7. -5.9 8. 4.9

9. -9.9 10. -2.7 11. 3.8 12. -12.7 13. $-\dfrac{4}{5}$ 14. $-\dfrac{1}{2}$ 15. $\dfrac{1}{24}$

16. $\dfrac{49}{60}$ 17. $-\dfrac{9}{14}$ 18. $\dfrac{7}{18}$ 19. 19.1 20. -18.9 21. -3.6 22. -3.8

23. -9.1 24. 8.6 25. $-\dfrac{2}{3}$ 26. $-\dfrac{21}{10}$ 27. $\dfrac{2}{5}$ 28. $-\dfrac{2}{7}$ 29. -16 30. $\dfrac{2}{7}$

31. -0.06 32. -7.13 33. 22.96 34. -8.06 35. -2.88 36. 27.28

37. $-\dfrac{14}{3}$ 38. $-\dfrac{3}{4}$ 39. 4 40. $-\dfrac{5}{6}$ 41. -1 42. $\dfrac{2}{3}$ 43. -7

44. -4 45. 5 46. -11 47. -1.6 48. -3

Extra Practice 24

1. -1 2. 5 3. -17 4. 3 5. -137 6. 3 7. 7 8. 7 9. -38

10. 36 11. 4 12. -32 13. -4 14. -16 15. $-\frac{5}{66}$ 16. $-\frac{14}{9}$

17. $-\frac{14}{19}$ 18. $\frac{4}{161}$ 19. $\frac{7}{8}$ 20. -16

Extra Practice 25

1. 7n - 28 2. 5x + 30 3. -2x + 14 4. -9y - 90 5. -4a - 12b

6. 20x - 30y 7. -5x - 10y + 30 8. 40x + 32y - 96 9. 18a - 9b + 27

10. 21p + 105q - 98 11. -18r + 45t - 63 12. -120a - 200b + 180

13. 6(x - 1) 14. 8(x + 3) 15. 4(x - 7) 16. 5(y - 6) 17. 7(x + 1)

18. 9(x - 7) 19. 8(6 - x) 20. 11(5 - x) 21. 3(2a + 3) 22. 7(2x - 7)

23. 5(2y + 3) 24. 6(3a - 5) 25. 10(5x - 7) 26. 8(4x + 3)

27. 6(x + 5 - 6a) 28. 15(x + 3y - 1) 29. b(x - 3y + 6) 30. a(x - 5 - y)

31. 3(2x + 3y - 8) 32. 5(5x - 3y + 15)

Extra Practice 26

1. 61 2. 194 3. -182 4. 32 5. -87 6. $-\frac{4}{3}$ 7. $\frac{1}{8}$ 8. $\frac{5}{6}$

9. -4 10. 8 11. $-6\frac{2}{5}$ 12. -144 13. $-\frac{8}{15}$ 14. 14 15. -18

16. 184 17. 3 18. 118 19. $\frac{1}{10}$ 20. $-2\frac{1}{2}$ 21. 62 22. -180

23. -1 24. $-\frac{2}{3}$ 25. $-\frac{11}{8}$ 26. -55 27. 1 28. 6 29. -16

30. -5 31. 7 32. 6 33. -8 34. -11 35. 3 36. 4

37. 77 38. -9 39. 2 40. -6 41. $\frac{5}{2}$ 42. -8 43. 1 44. 6

45. 2 46. $-\frac{4}{9}$ 47. $-\frac{15}{2}$ 48. -36 49. 75 50. -66 51. 0

52. 0 53. 0 54. 0

1. 8   2. -30   3. 9 cm, 17 cm   4. 23 m, 16 m   5. 48, 50, 52

6. 85, 87, 89   7. 2 ft, 6 ft, 12 ft   8. 50 m, 100 m, 300 m

9. 20°, 60°, 100°   10. 46°, 92°, 42°   11. $214   12. 20

1. (a) Answers may vary: (b) 49   2. d = 8   3. (a) 162, 486, 1458, 4374, 13122, 39366, 118098; (b) 19th   4. 7, 9, 63   5. a = 8, b = 4
6. $5 + 7 \cdot 9 - 3 = 65$, or $5 + 9 \cdot 7 - 3 = 65$; the commutative law of multiplication.   7. $4 + 8^2 \cdot 6 - 2 = 386$, because $a^2 \cdot b \neq b^2 \cdot a$, although students might phrase this orally and not symbolically.

8. $31 \cdot (87 - 19) = 2108$ 9. $2^9 \div 8^2 \cdot 3^2$.

Extended Synthesis Exercises

1. 7 days   2. 60   3. (a) $38 = 2 \cdot 19$; (b) $0 = 2 \cdot 0$; (c) $47 = 2 \cdot 23 + 1$; (d) The sum of two even numbers is even. (e) The sum of two odd numbers is even. (f) The sum of an even number and an odd number is odd. (g) The product of an even and an odd number is even. (h) It is odd. The product of an odd number and another odd number is odd, so a^2 is odd. The product of an odd number and an even number is even, so $a \cdot b$ is even. The sum of an even number and an odd number is odd, so $a^2 + a \cdot b$ is odd.   4. 11,011; 5115
5. 7 chickens and 15 pigs   6. Write word names for each number; then use the alphabetical order to arrange the numbers.   7. 1, 4, 9, 121, 484, 676, 10201, 12321, 14641, 40804, 44944, 69696, 94249; not necessarily   8. 1, 8, 343, 1331, 1030301, 1367631   9. 64   10. 169   11. The nth square number is the sum of the first n odd numbers.   12. 29th   13. There are 16 addends of odd numbers used to get a square number. Thus the sum is 16^2, or 256.
14. 333 is a palindrome, $3 + 3 + 3 = 9$, which is a square, and $3 \cdot 3 \cdot 27$, which is a cube. Other answers are 202 and 808.

Exercises for Thinking and Writing

1. In a "take away" problem situation, we start with a certain quantity, take away part of it, and find how much remains. In a "how much more" problem situation, we start with a certain quantity and find how much more must be added to it in order to obtain a larger quantity.   2. (a) In a division problem, rounding the divisor helps to estimate the number of digits in the quotient. (b) Rounding and estimating is also a quick way to check a computation. (c) Rounding and estimating is a good way to approximate the result of a computation when it is not convenient to perform the actual computation.   3. (a) A person hired to work at h dollars per hour works for

n hours. Multiplication is used to find the total pay, *h* • *n* dollars.
Answers may vary. **(b)** Tickets for a concert cost *d* dollars each. You are
buying *n* tickets. Multiplication is used to find the total price,
d • *n* dollars. Answers may vary. **4. (a)** A group of *p* people rents a car
for *d* dollars. Division is used to find the cost per person, $\frac{d}{p}$ dollars.
(b) Items in a store are priced at *n* items for *d* dollars. Division is used to
find the cost of one item, $\frac{d}{n}$ dollars. **5.** Tickets for a concert cost $29 for
adults and $3 for children. What is the total cost of admittance for 7 adults
and 1 child? Answers may vary.

Chapter 2, P. 125
Calculator Connection

1. a = 6558 **2.** a = 25,588, b = 33,699 **3.** a = 7889, b = 9887 **4.** Primes are
421 and 587

Extended Synthesis Exercises

1. Add on the left sides; remove a factor equal to 1 on the right sides. The
pattern is $1 + 2 + 3 + \cdots + n = \frac{n \cdot (n + 1)}{2}$, where the dots stand for the
symbols we did not write. **2.** 21 **3.** 36 **4.** 55 **5.** 5050 **6.** 36 **7.** 91
8. 20,100 **9.** The 16th **10.** Yes, it is the 28th. Note that $\frac{28 \cdot 29}{2} = 406$.
11. No. There is no whole number *n* for which $\frac{n \cdot (n + 1)}{2} = 423$. To explain
another way, note that $\frac{28 \cdot 29}{2} = 406$ and $\frac{29 \cdot 30}{2} = 435$, but 423 is between 406
and 435. **12.** 3, 4; 7, 8; 28, 29; 100, 101 **13.** Every pair of consecutive
numbers is an even number and an odd number. **14.** To find the *n*th triangular
number, multiply the consecutive numbers *n* and *n* + 1, and divide by 2.
15. Palindrome primes: 13, 11, 101 **16.** 3 and 5; 5 and 7; 11 and 13;
17 and 19; 29 and 31; 41 and 43; 59 and 61; 71 and 73; 101 and 103; 107 and
109; 137 and 139; 149 and 151; 179 and 181; 191 and 193; 197 and 199.
17. 6 = 3 + 3; 8 = 3 + 5; 10 = 5 + 5; 12 = 5 + 7; 14 = 7 + 7; 16 = 5 + 11;
18 = 11 + 7; 20 = 17 + 3; 22 = 11 + 11; 24 = 13 + 11; 26 = 13 + 13;
28 = 11 + 17; 30 = 19 + 11; 32 = 19 + 13; 34 = 17 + 17; 36 = 19 + 17;
38 = 19 + 19; 40 = 29 + 11; 42 = 29 + 13; 44 = 37 + 7; 46 = 23 + 23;
48 = 37 + 11. **18.** $\frac{32}{3125}$, $\frac{64}{15,625}$, $\frac{128}{78,125}$, $\frac{256}{390,625}$, $\frac{512}{1,953,125}$.
19. (a) $\frac{73,553}{138,346}$; **(b)** $\frac{28,372}{138,346}$; **(c)** $\frac{36,421}{138,346}$

Exercises for Thinking and Writing

1. To simplify when using fractional notation, we first factor the numerator and factor the denominator. Then we remove a factor of 1. That is, we remove pairs of factors that are common to the numerator and the denominator. **2.** The tests of divisibility are helpful in factoring the numerator and the denominator of a fraction. The factors common to the numerator and the denominator of a fraction are easily determined when the numerator and the denominator are factored into prime factors. This is helpful in simplifying fractions. **3.** Find the area of a small rectangular object that has a length of $\frac{9}{10}$ in. and a width of $\frac{7}{8}$ in. **4.** Use computational examples: $\frac{1}{2} \cdot 24 = 12$, but $24 \div \frac{1}{2} = 24 \cdot 2 = 48$. **5.** 1993, 1997, 1999, and 2003. Consult a larger table of primes than is in the text, or divide by numbers smaller than the number. **6.** Answers may vary. Good sources are books on the history of mathematics.

Chapter 3, P. 181
Calculator Connection

1. $a = 2$, $b = 8$ **2.** The largest is $\frac{4}{3} + \frac{5}{2} = \frac{23}{6}$. **3.** The largest is $4 + \frac{6}{3} \cdot 5 = 14$ or $4 + \frac{5}{3} \cdot 6 = 14$. **4.** $\frac{3}{4}, \frac{7}{9}, \frac{17}{21}, \frac{19}{22}, \frac{13}{15}, \frac{15}{17}, \frac{13}{12}$.

Extended Synthesis Exercises

1. (a) $\frac{1}{2}, \frac{2}{3}, \frac{3}{4}, \frac{4}{5}$; **(b)** $\frac{9}{10}$ **2. (a)** $\frac{1}{2}, \frac{3}{4}, \frac{7}{8}, \frac{15}{16}$; **(b)** 1 **3.** The length of their act is the LCM of 6 and 4, which is 12 min. **4. (a)** 24, 48, 72, 96, and so on; **(b)** 24, which is the LCM of 6 and 8. **5.** $r = \frac{600}{13}$ **6. (a)** 0; **(b)** 1; **(c)** 0 **(d)** 1; **(e)** $\frac{1}{2}$; **(f)** 1; **(g)** $\frac{1}{2}$; **(h)** 0; **(i)** 0; **(j)** $\frac{1}{2}$; **(k)** 1; **(l)** $\frac{1}{2}$ **7. (a)** 6; **(b)** 5; **(c)** 12; **(d)** 18; **(e)** 19; **(f)** 13; **(g)** 15; **(h)** 100 **8. (a)** 8; **(b)** 12; **(c)** 14; **(d)** 28; **(e)** 13; **(f)** 7; **(g)** 16; **(h)** 97 **9. (a)** 3; **(b)** 1; **(c)** 13; **(d)** $26\frac{1}{2}$ **10. (a)** 2; **(b)** $\frac{1}{2}$; **(c)** 2; **(d)** 3; **(e)** 0; **(f)** 3

Exercises for Thinking and Writing

1. It is not correct to multiply the whole numbers and then the fractions. The mixed numerals should be converted to fractional notation first and then multiplied. $2\frac{1}{4} \cdot 3\frac{2}{5} = \frac{9}{4} \cdot \frac{17}{5} = \frac{153}{20} = 7\frac{13}{20}$ **2.** When multiplying a whole number and a mixed numeral, it is not correct to multiply the product of the whole numbers and the product of the whole number and the fractional part of the mixed numeral. The mixed numeral should be converted to the fractional notation first, and then the multiplication can be done. $5 \cdot 3\frac{2}{5} = \frac{5}{1} \cdot \frac{17}{5} = \frac{5 \cdot 17}{1 \cdot 5} = \frac{5}{5} \cdot \frac{17}{1} = 17$ This is the equivalent of the following: $5 \cdot 3\frac{2}{5} = 5\left(3 + \frac{2}{5}\right) = 5 \cdot 3 + 5 \cdot \frac{2}{5} = 15 + 2 = 17$. **3.** Least common multiples are used in adding and subtracting with fractional notation when denominators are different. The least common multiple of the denominators is the least common denominator. We multiply by 1, using an appropriate notation n/n to obtain the LCD for each number. Then we add or subtract and simplify, if possible.

4. An investor bought two shares of stock at $\$15\frac{3}{4}$ and two shares of another stock at $\$28\frac{5}{8}$. Find the total cost of the purchase.

Chapter 4, P. 215
Calculator Connection

1. $a = 2$, $b = 8$, $c = 3$, $d = 6$ 2. $\$66.70$, $\$77.82$, $\$88.94$, $\$100.06$, $\$111.18$, $\$122.30$, $\$133.42$ 3. $\$2029.66$, $\$1950.88$, $\$1872.10$, $\$1793.32$ 4. $a = 5$, $b = 9$
5. Magic sum = 7.8; first row: 4.55, 1.3; second row: 0, 2.6, 5.2   6. First row: 0.81; second row: 0.27, 1.35, 2.43; third row: 1.89   7. First row: 43.12; second row: 9.52, 39.76; third row: 15.12, 6.72; fourth row: 10.64, 11.76

Extended Synthesis Exercises

1. The percentage of bone weight that is removed. Check with local groceries or consumer buyer guides or your local county agent.   2. (a) False; $\frac{1}{4} + \frac{1}{4} = \frac{1}{2}$, and $\frac{1}{2} < 1$; (b) False; $\frac{3}{4} + \frac{1}{4} = 1$; (c) True   3. 744.16 should be 744.17; 764.65 should be 723.68; 848.65 should be 808.68; 801.05 should be 760.08; 533.09 should be 492.13

Exercises for Thinking and Writing

1. The decimal points were not lined up.   2. The decimal points were not lined up.   3. Decimal notation can be defined in terms of fractional notation as a whole number plus a sum of fractions, each with a power of ten as a denominator. For example, 12.4598 can be defined as $12 + \frac{4}{10} + \frac{5}{100} + \frac{9}{1000} + \frac{8}{10,000}$.   4. A business has $\$23,456.76$ in its corporate checking account. It makes a deposit of $\$7655.68$. Then it has expenses of $\$67.95$ and $\$123.69$. Answers may vary.

Chapter 5, P. 263
Calculator Connection

1. $0.\overline{142857}$ 2. $0.\overline{285714}$ 3. $0.\overline{428571}$ 4. $0.\overline{571428}$ 5. $0.\overline{714285}$
6. $0.\overline{857142}$ 7. $0.\overline{1}$ 8. $0.\overline{01}$ 9. $0.\overline{001}$ 10. $0.\overline{0001}$ 11. $0.\overline{012345679}$
12. $0.\overline{123456790}$ 13. $1.\overline{234567901}$ 14. $12.\overline{345679012}$ 15. $123.\overline{456790123}$
16. $\frac{2}{3}$, $\frac{5}{7}$, $\frac{15}{19}$, $\frac{11}{13}$, $\frac{17}{20}$, $\frac{13}{15}$ 17. $2.56 \cdot 6.4 \div 51.2 - 17.4 + 89.7 = 72.62$
18. $(0.37 + 18.78) \cdot 2^{13} = 156,876.8$

Extended Synthesis Exercises

1. b - 1. There are b numbers less than b, which can be remainders when division is used to find decimal notation. If 0 is a remainder, the decimal notation terminates. Then there are b - 1 nonzero remainders. After each one occurs, one of the remainders must reoccur. Thus there can be no more than b - 1 digits in the repeating part of the decimal. **2.** 14, 23 **3.** 0.5, 0.3 **4.** $1.44 **5.** Now $\frac{1}{3} + \frac{2}{3}$ = 0.33333333. . . + 0.66666666. . . = 0.99999999. . . Therefore, 1 = 0.99999999. . . because $\frac{1}{3} + \frac{2}{3}$ = 1. **6.** 2 = $1.\overline{9}$ **7. (a)** Always; **(b)** never; **(c)** sometimes; **(d)** always

Exercises for Thinking and Writing

1. Since the denominator is 61, a prime number other than 2 or 5, we cannot multiply to get a denominator that is a power of 10. Thus we can divide to find decimal notation. The first nine decimal places are 0.721311475. Decimal notation will terminate or repeat, but it is too long to be found on most calculators. There are actually 60 digits in the repeating part of the decimal notation, so the long division will be quite lengthy. The answer is 0.721311475 409836065 573770491 803278688 524590163 934426229 508196. **2.** Estimating can be used before a computation is done to get an idea of the answer. For example, we know that the answer to 7.2 x 8.9 is about 7 x 9, or 63. Thus the answer will be "two digits, a decimal point, followed by some digits". Thus we multiply 72 and 89 and place the decimal point two digits from the left. **3.** The computation in Section 5.1, where decimal notation is converted to fractional notation, justifies multiplication with decimal notation. **4.** Rudy buys 8 baseball jackets at $79.95 each and 3 portable cassette players at $149.59 each. How much does he spend in all? Answers may vary.

Chapter 6, P. 299
Calculator Connection

1. 8188.068 **2.** 3681.437 **3.** 1050 **4.** 129.455 **5.** x = 312.48, y = 466.2 **6.** x = 43.786, z = 40.295 **7.** y = 0.352 **8.** x = 4258.5, z = 10,094.267

Extended Synthesis Exercises

1. New York 112, Portland 92 **2.** 5888 **3.** 105 min **4.** New York **5.** Vermont **6.** $613.8 million **7.** $3.6042 billion **8.** 0.1706 **9.** $461.3 million **10.** $74.3 million **11.** $13.67 per person **12.** $34.81 per person **13.** $45.08 per person **14.** Illinois

Exercises for Thinking and Writing

1. A *ratio* is a quotient of two quantities. A *rate* is a ratio that is used to compare two different kinds of measures. A *proportion* is an equation that states that two pairs of numbers have the same ratio. **2.** A unit price is a rate because it is a ratio that compares two different kinds of measure, price and number of units. **3.** It costs $128.95 to buy 18 cassette tapes. How many cassette tapes can be purchased for $789.89?

Chapter 7, P. 361
Calculator Connection

1. (a) 253; **(b)** 8382.888; **(c)** $487.20 **2.** 50.04% **3.** 49.90% **4.** 26.30%
5. 0.00467105 **6.** $45,137.69 **7.** $19,381.20; $20,195.21; $21,043.41; $21,927.23; $22,848.17; 22.84% **8.** 19.48% **9.** 0.0000006%

Extended Synthesis Exercises

1. (a) #1 pays $\frac{1}{5} \cdot \frac{1}{5}$,

#2 pays $\frac{1}{5} \cdot \frac{1}{5} + \frac{1}{4} \cdot \frac{1}{5}$,

#3 pays $\frac{1}{5} \cdot \frac{1}{5} + \frac{1}{4} \cdot \frac{1}{5} + \frac{1}{3} \cdot \frac{1}{5}$,

#4 pays $\frac{1}{5} \cdot \frac{1}{5} + \frac{1}{4} \cdot \frac{1}{5} + \frac{1}{3} \cdot \frac{1}{5} + \frac{1}{2} \cdot \frac{1}{5}$,

#5 pays $\frac{1}{5} \cdot \frac{1}{5} + \frac{1}{4} \cdot \frac{1}{5} + \frac{1}{3} \cdot \frac{1}{5} + \frac{1}{2} \cdot \frac{1}{5} + \frac{1}{5}$;

(b) #1 pays 4%, #2 pays 9%, #3 pays $15\frac{2}{3}$%, #4 pays $25\frac{2}{3}$%, #5 pays $45\frac{2}{3}$%;
(c) 87% **2.** $10,560; 5.6% **3.** $33\frac{1}{3}$%, **4.** 40%. Sucessive 20% discounts means that you pay 80% of 80%, or 64%, which is only a 36% discount.
5. 7 **6.** Answers depend on local data. **7.** He bought the plaques for $166\frac{2}{3}$ + $250, or $416\frac{2}{3}$, and sold them for $400, so he lost money. **8.** $66\frac{2}{3}$% **9.** 26%
10. 71% **11.** 90% **12.** 70% **13.** 345% **14.** 98% **15.** 2.5% **16.** 0%
17. 5.29% **18.** 6.4% **19.** 50%, 25%, 12.5%, 6.25%, 3.125%, 1.5625%
20. (a) $8.10; **(b)** $9.50; **(c)** 15%x = (10% + 5%)x = 10% $\cdot$ x + 5% $\cdot$ x, by the distributive law. **21. (a)** $524.88; 1.08; x + 8% $\cdot$ x = 1 $\cdot$ x + 8% $\cdot$ x = 1 $\cdot$ x + 0.08 $\cdot$ x = (1 + 0.08) $\cdot$ x = 1.08 $\cdot$ x, by the distributive law; **(b)** $413.10; 0.85; x - 15% $\cdot$ x = 1 $\cdot$ x - 0.15 $\cdot$ x = (1 - 0.15) $\cdot$ x = 0.85 $\cdot$ x, by the distributive law. **22. (a)** Melissa was 13 mph over the speed limit. Emmett was 12 mph over the speed limit. Police usually fine depending on the number of miles per hour over the speed limit. but 13 is a 20% increase over 65 mph, and 12 is 30% increase over 40 mph.

(b) 2 mph is only 1 mph over the speed limit, but it is double the limit, or a 100% increase. 109 is 54 mph over the limit, but it is about a 98% increase. Which is "worse" is a matter of debate.

Exercises for Thinking and Writing

1. Daily uses of percent include statistics we read in newspapers and magazines (e.g., crime rates, population increase or decrease, and many others), sales tax, price reductions or increases, and interest rates. **2.** In solving percent problems using equations, we translate to one of the three different types of equations: (1) *What* is some percent of a base amount? (2) An amount is *what percent* of a base amount? (3) An amount is some percent of *what base amount*? In each case, we solve for the unkown quantity, given the other two quantities. In solving percent problems using proportions, we translate to a proportion of the form "amount is to base as percent number is to 100," or $\frac{\text{Amount}}{\text{Base}} = \frac{\text{Percent number}}{100}$. In each case, in addition to 100, we know two of the other three quantities (amount, base, percent number) and solve for the third. **3.** Percent increase or decrease tells what percent of an original amount an amount of increase or decrease is. A common error is to use the increased or decreased amount (that is, the original amount plus the amount of increase or decrease) in place of the original amount. **4.** A real estate commission on the sale of a house is 6.7%. The house sells for $158,600. The commission is $10,626.20. Answers may vary.

Chapter 8, P. 403
Extended Synthesis Exercises

1. $5.2 billion; answers may vary. For the actual revenue in 1995, check with a stockbroker. Then make a prediction about the year 2000.

2.

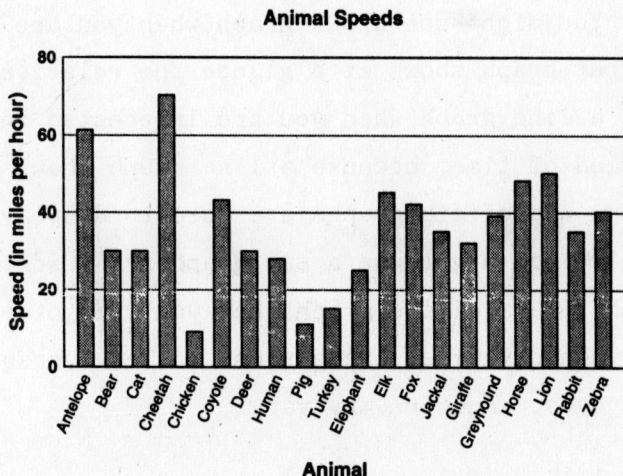

Animal Speeds

3. 35.9 mph   **4.** 36.3 mph; animal speed is 8.3 mph higher than average human speed.   **5.** Below 20 mph: chicken, pig, turkey (all grain and grass eaters); 20 - 30 mph: bear, cat, deer, human, elephant; 31 - 40 mph: jackal, giraffe, greyhound, rabbit, zebra (all four-legged; some feed on the others); 41 - 50 mph: coyote, elk, fox, horse, lion (all four-legged; some feed on the others); above 50 mph: antelope, cheetah (both four-legged; cheetah may feed on antelope).   **6.** Antelope and cheetah; both above 60 mph in speed. Cheetah may feed on the antelope.   **7.** Meat-eaters: 45 mph; both: 29 mph; grains, grasses: 31.9 mph. The meat-eaters have a higher average speed, probably because they must pursue their prey.   **8.** Two-legged: $17\frac{1}{3}$ mph; four-legged: 39.18 mph. The four-legged are more than twice as fast. Answers to Exercises **9 - 17** may vary due to estimating error in reading data from the graph and possible interpretations of the graph.   **9.** About $58 per day   **10.** About $60 per day **11.** About $43 per day   **12.** 1989   **13.** Competition between companies and lower gasoline prices. Answers may vary.   **14.** 1989 through 1990 **15.** Around April (one-third) of 1991   **16.** $61 hotel--slow increase; $54 car rental--decrease; $44 food   **17.** 1995 **18.** Answers will vary and depend on the stock chosen.

Exercises for Thinking and Writing

1. The average, median, and mode are "center points" that characterize a set of data. You might use the average to find a center point that is approximately midway between the two extreme values of the data. The median might be used to find a center point that is the middle of all the given data values. That is there are as many values greater than the median than there are less than the median. The mode might be used as a center point that represents a values(s) that occurs most frequently.   **2.** Bar and line graphs are used to display data. You might use a bar graph when you are interested in comparisons, because a bar graph shows at a glance the relative size of quantities. You might use a line graph when you are interested in changes that take place over a period of time, because a line graph shows patterns or trends at a glance.   **3.** Bar and circle graphs are used to display data. The circumstances under which you might use a bar graph are described in Exercise 2 above. You might use a circle graph when you are interested in the breakdown of a quantity into various categories. **4.** The average of a person's salary over four years. Answers may vary.

Chapter 9, P. 463
Calculator Connection

1. Answers may vary.   2. 1417.989 in.   3. 160,005.908 in^2 4. 1729.270 in^2
5. 125,663.706 ft^2 6. Diameter = 18 in., radius = 9 in., circumference =
56.549 in.   7. Radius = 6.088 in.; diameter = 12.176 in.; area = 116.427 in^2

Extended Synthesis Exercises

1. (a) 35.2 ft^2, (b) 4   2. Divide by 4.   3. The times, in seconds, are 418,
406, 396, and 388.   If the pattern continues, the times, in seconds, are 382,
378, 376, 376, and so on.   The times eventually bottom out to 376 seconds by
the seventh week, which she cannot improve.   4. They grow $\frac{1}{6}$ in. per month.
He will trim $2\frac{2}{3}$ times per month, or 32 times per year.   5. 100 ft^2 6. 2 ft^2
7. 1.875 ft^2 8. $4\frac{1}{3}$ ft^2 9. $13\frac{1}{3}$ ft^2 10. 7.83998704 m^2 11. 42.05915 cm^2
12. 142.75 ft^2, about 99.13%   13. 1701; 42%

Exercises for Thinking and Writing

1. Answers might include the following: (1) The metric system is used in most
countries.   (2) It is easier to convert from one metric unit to another,
because the metric system is based on the number 10.   2. Answers might include
the following: (1) We are already familiar with the American system.
(2) Converting to the use of the metric system would be a large task,
considering the many uses of American measures in our daily lives.   3. The
metric system was adopted by law in France about 1790 during the rule of
Napoleon I.

Chapter 10, P. 501
Calculator Connection

1. About 314.159 metric tons

Extended Synthesis Exercises

1. 0.65 ft^3 2. 1.342 ft^3 3. 0.033 ft^3 4. 94,200 ft^3 5. 95.426 in^3, or

0.055 ft^3 6. 11.454 in^3, or 0.0066 ft^3 7. $386\frac{2}{3}$ ft^3; 14.32 yd^3
8. 1204.260429 ft^3; about 22,079,692.8 ft, or 4181.76 mi   9. (a) 36.744
square miles; (b) The state of Indiana has a land area of about 36,185 square
miles. (c) 5.1218×10^{11} cubic feet   10. An ounce of pennies weighs 1 oz; an
ounce (capacity) of water weighs (8.3453 x 16)/128 oz, or 1.0432 oz.   Thus the
water weighs more.   11. $188.40   12. About 34%   13. $101.25

Exercises for Thinking and Writing

1. Gabriel Daniel Fahrenheit, a Dutch instrument maker, used the idea of calibrating a thermometer to the melting point of ice and the heat of human blood. In 1724, he constructed a thermometer in which he fixed 32° as the freezing point of water and 96° as normal body temperature. On this scale, which was named for Fahrenheit, water boiled at 212°, but this temperature became the upper fixed point only after Fahrenheit's death.

In 1742, Anders Celsius, a Swedish astronomer, devised a temperature scale in which the interval between the freezing and boiling points of water was divided into 100 degrees. Celsius originally set the freezing point of water at 100° and the boiling point at 0°. In 1745, his friend Carl Linnaeus inverted the scale to give us the now-familiar centigrade scale, generally known as the Celsius scale. 2. See the volume formulas in the Summary and Review at the end of the chapter.

Chapter 11, P. 539
Calculator Connection

1. The largest value is 8 for either $6 \div 4 \cdot 8 - 2^2$, or $8 \div 4 \cdot 6 - 2^2$.
2. The largest value is 34 for $6 - 4 + 8^2 \div 2$.   3. $-32 \cdot (88 - 29) = -1888$
4. $3^5 \div 10^2 - 5^2 = -22.57$ 5. You get the same number. If the number is a, then $1/a$ is its reciprocal, and the reciprocal of $1/a$ is $a/1$, or a.
6. **(a)** 52, 52, 28.130169; **(b)** -24, -24, -108.307025

Extended Synthesis Exercises

1. For $x < 0$ 2. The dimensions 7 by 7 give the largest area.

LENGTH	WIDTH	AREA
7	7	49
8	6	48
9	5	45
7.2	6.8	48.96
8.1	5.9	47.79
10	4	40

3. 403 and 397   4. 12 and -7 5. 68   6. **(a)** -4, -9, -15; **(b)** -2, -6, -10; **(c)** -18, -24, -31; **(d)** $-\frac{1}{4}$, $\frac{1}{8}$, $-\frac{1}{16}$ 7. **(a)** $-7 + (-6) + (-5) + (-4) + (-3) + (-2) + (-1) + 0 + 1 + 2 + 3 + 4 + 5 + 6 + 7 + 8 = 8$; **(b)** 0   8. -1 9. -1; Consider reciprocals and pairs of products of negative numbers.   10. $-\frac{24}{7}$

11. The expressions are not equivalent. For example, $|-4 + 7| = 3$ and $|-4| + |7| = 11$. **12.** 5.8×10^8 mi **13.** -83; that is, total attendance is down 498 people, for an average decrease of 83. **14.** Up 15 points

Exercises for Thinking and Writing

1. Three examples are $\frac{6}{13}$, -23.8 and $\frac{43}{5}$. Answers may vary. **2.** Yes; the numbers 1 and -1 are their own reciprocals. If you multiply 1 by 1, you get 1. If you multiply -1 by -1, you get 1. **3.** We know that $a + (-a) = 0$, so the opposite of $-a$ is a. That is, $-(-a) = a$.

Chapter 12, P. 583
Calculator Connection

1. 2233.526854 **2.** 0.572 **3.** 52, 52, 24.884 **4.** -24, -24, -9.569 **5.** 9, -9, 9, -9; $x^2 = (-x)^2$, $-1 \cdot x^2 = -x^2$

Extended Synthesis Exercises

1. $\frac{5abd}{cd}$, $\frac{5abc}{c^2}$; answers may vary. **2.** They are not equivalent. Let $a = 2$ and $b = 3$. Then $(a + b)^2 = (2 + 3)^2 = 5^2 = 25$ and $a^2 + b^2 = 2^2 + 3^2 = 4 + 9 = 13$.

Thus, $(a + b)^2 \neq a^2 + b^2$. **3.** $-15x = 8$; answers may vary. **4.** $x - \frac{1}{5} = -\frac{11}{15}$; answers may vary. **5.** $15x + 7 = -1$; answers may vary. **6.** $x^2 = 25$, or $|x| = 5$; answers may vary. **7.** $|x| = -5$; or $0 \cdot x = 5$; answers may vary. **8.** The top section is 20 ft, the fuel section is 100 ft, and the rocket section is 120 ft. **9.** $1.50 **10.** 72 **11.** 72 **12.** 9, 15, 4, 36 **13.** $a^2 - b^2$ **14.** x^2

Exercises for Thinking and Writing

1.
$$4 - 3x = 5$$
$$3x = 9 \qquad (1)$$
$$x = 3 \qquad (2)$$

(1) 4 should have been subtracted, not added, on the right side. Also, the minus sign preceding 3x has been dropped. (2) This step would give the correct result if the previous step were correct. The correct steps are

$$4 - 3x = 5$$
$$- 3x = 1 \qquad (1)$$
$$x = -\frac{1}{3}. \qquad (2)$$

2.
$$2x - 5 = 7 \qquad (1)$$
$$2x = 12 \qquad (2)$$
$$x = 6 \qquad (3)$$

(1) Using the distributive law to remove parentheses, x was multiplied by 2, but 5 was not. (2) and (3) These steps would give the correct result if the previous step were correct. The correct steps are

$$2x - 10 = 7 \qquad (1)$$
$$2x = 17 \qquad (2)$$
$$x = \frac{17}{2}. \qquad (3)$$

3. The distributive laws are used to multiply, factor, and collect like terms in this chapter. 4. First, $(b + c)a = a(b + c)$, by the commutative law of multiplication. Then $a(b + c) = ab + ac$, by the distributive law. Finally, using another application of the commutative law, we have $ab + ac = ba + ca$, so $(b + c)a = ba + ca$. 5. An expression is a collection of mathematical symbols that denotes a number when the variables are replaced by numbers; for example, $x + 7$, $8y$, $3x - 8y$, and so on. An expression does not contain an equality symbol "=." An equation contains an equality symbol "=." It has expressions on either side of the = symbol. An equation asserts that the expressions on either side of the equals sign represent the same number.

6. A subtraction principle would assert that if $a = b$ is true, then we can subtract any number c on both sides and get another true equation $a - c = b - c$. But we already know that $a + (-c) = b + (-c)$, by the addition principle, and $a + (-c) = a - c$ and $b + (-c) = b - c$. A division principle would assert that if $a = b$ is true, then we can divide by any nonzero number c on both sides and get another true equation $a \div c = b \div c$. But we already know that $a \cdot \frac{1}{c} = b \cdot \frac{1}{c}$, by the multiplication principle, and $a \cdot \frac{1}{c} = a \div c$ and $b \cdot \frac{1}{c} = b \div c$. 7. Suppose we have a particular formula that expresses a letter, say L, in terms of another letter, say a. In an application, you frequently have values of L given and want to compute values of a. It would be faster to do this if you first solved for a in terms of L.

TEST AID: NUMBER LINES

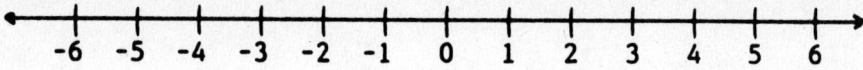

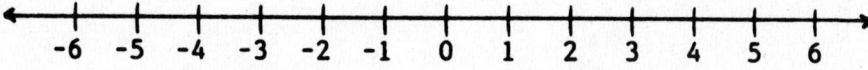

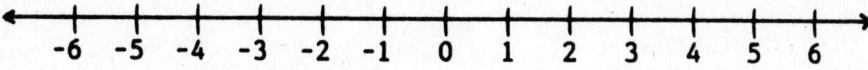

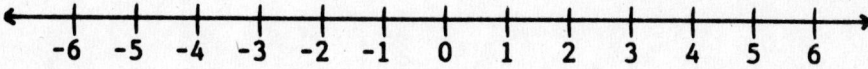

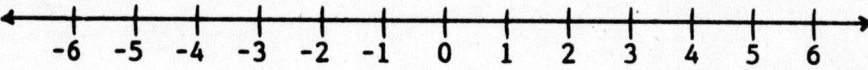

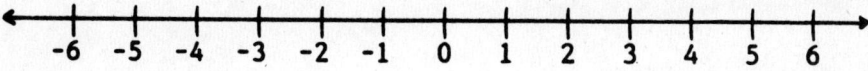

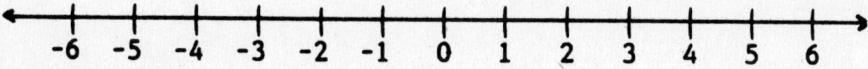

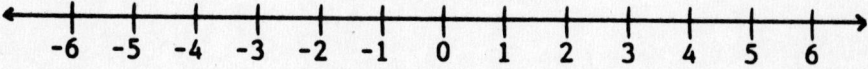

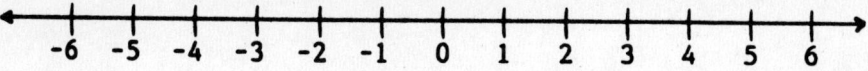

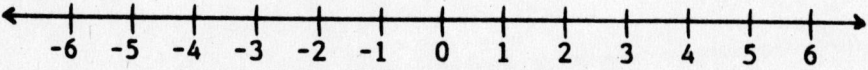

TEST AID: RECTANGULAR COORDINATE GRIDS

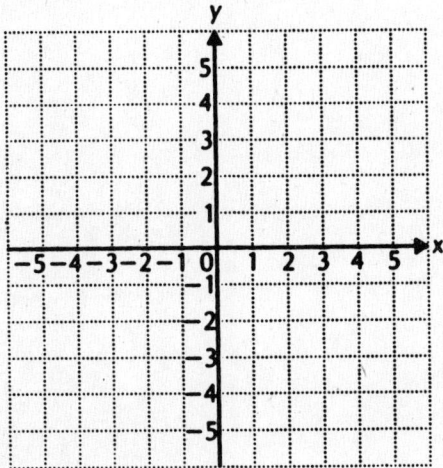

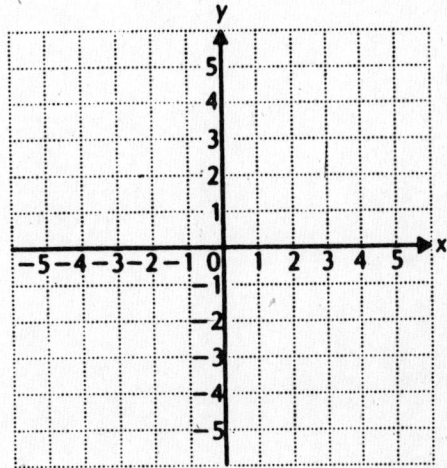

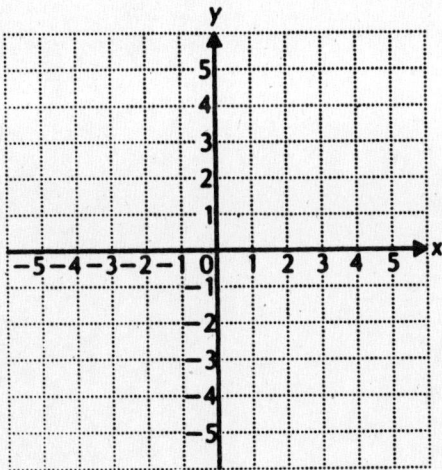

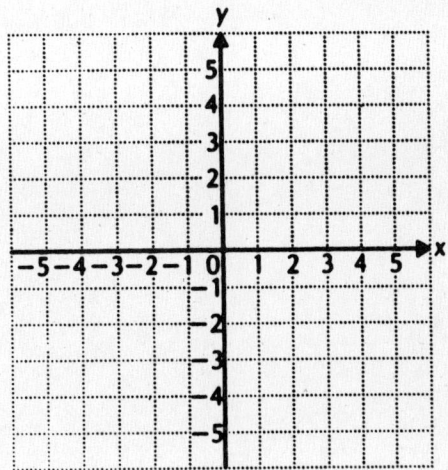

TRANSPARENCY MASTER: NUMBER LINES

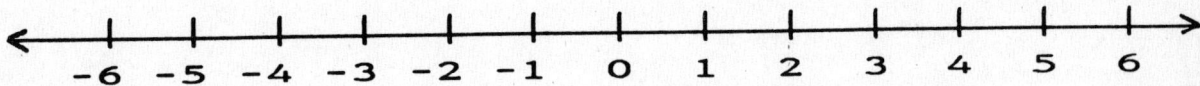

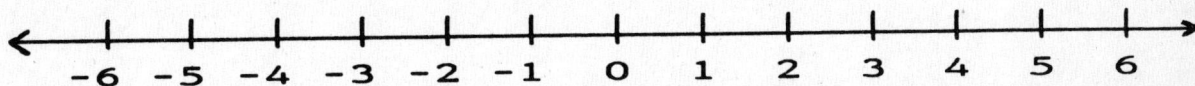

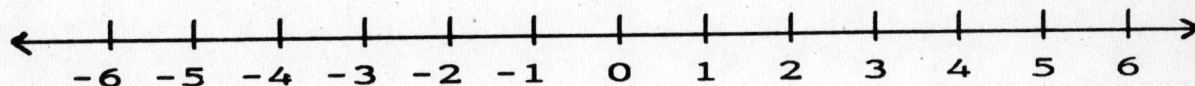

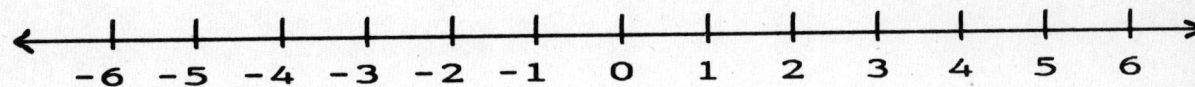

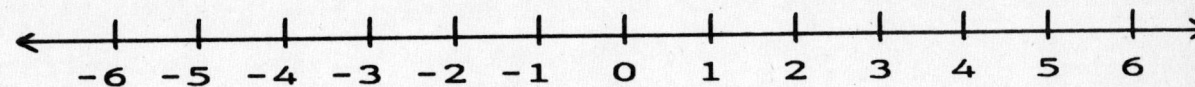

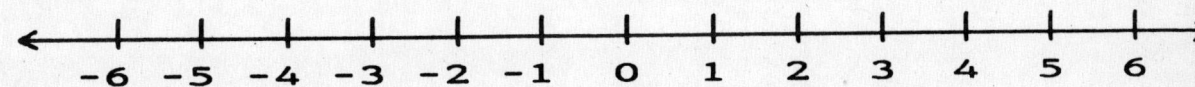

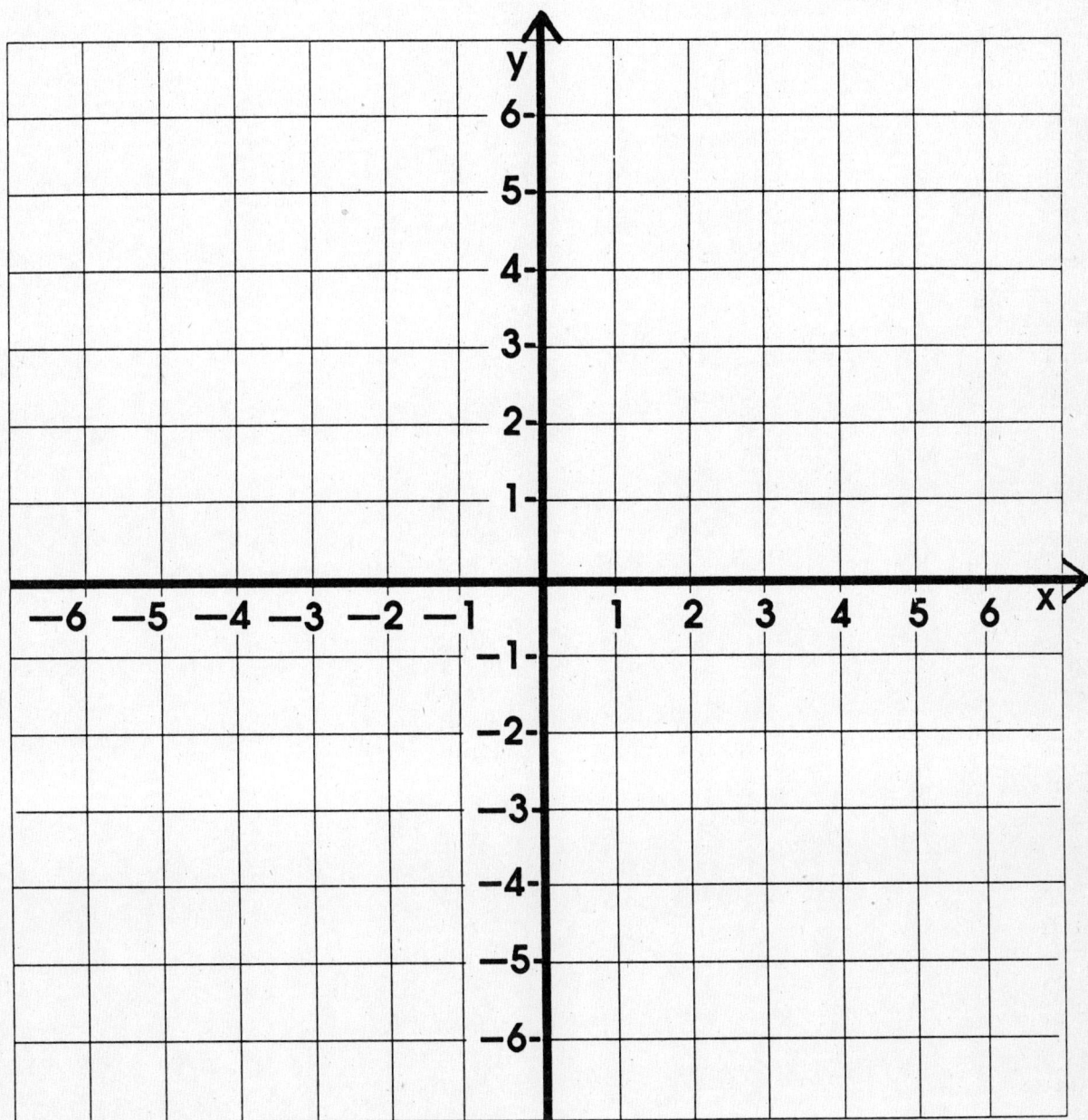

The Bittinger/Keedy System of Instruction

CONVERSION GUIDE

BASIC MATHEMATICS, SEVENTH EDITION
Bittinger/Keedy

This conversion guide is designed to help you adapt your syllabus for Keedy/Bittinger, *Basic Mathematics, Sixth Edition* to Bittinger/Keedy, *Basic Mathematics, Seventh Edition* by providing a Section-by-Section reference from the Sixth to the Seventh Edition. Many minor revisions and refinements have been made in additon to the changes specified here.

Chapter 1 - 6th and 7th - Operation on the Whole Numbers

Sec.	Basic Mathematics 6th	Sec.	Basic Mathematics 7th
1.1	Standard Notation	1.1	Standard Notation
1.2	Addition	1.2	Addition
1.3	Subtraction	1.3	Subtraction
1.4	Rounding and Estimating; Order	1.4	Rounding and Estimating; Order
1.5	Multiplication	1.5	Multiplication
1.6	Division	1.6	Division
1.7	Solving Equations	1.7	Solving Equations
1.8	Solving Problems	1.8	Solving Problems
1.9	Exponential Notation & Order of Operations	1.9	Exponential Notation & Order of Operations

Comments: In Section *1.2a* perimeter is introduced. In Section *1.9c* averaging of whole numbers is presented.

Chapter 2 - 6th and 7th - Multiplication and Division: Fractional Notation

Sec.	Basic Mathematics 6th	Sec.	Basic Mathematics 7th
2.1	Factorizations	2.1	Factorizations
2.2	Divisibility	2.2	Divisibility
2.3	Fractions	2.3	Fractions
2.4	Multiplication	2.4	Multiplication
2.5	Simplifying	2.5	Simplifying
2.6	Multiplying and Simplifying	2.6	Multiplying and Simplifying
2.7	Reciprocals and Division	2.7	Reciprocals and Division

Comments: In Section *2.2* the order in which divisibility rules appear is new. In the Sixth Edition it is 2, 9, 3, 4, 8, 6, 10, 5 and in the Seventh Edition it goes 2, 3, 6, 9, 10, 5, 4, 8. In Section *2.3*, the exercises drilling 15/0 (division not defined) are now in the margins, exercise set, summary and review, and chapter test. This concept was included in the Sixth Edition but not drilled.

Chapter 3 - 6th and 7th - Addition and Subtraction: Fractional Notation

Sec.	Basic Mathematics 6th	Sec.	Basic Mathematics 7th
3.1	Least Common Multiples	3.1	Least Common Multiples
3.2	Addition	3.2	Addition
3.3	Subtraction and Order	3.3	Subtraction and Order
3.4	Mixed Numerals	3.4	Mixed Numerals
3.5	Addition and Subraction Using Mixed Numerals	3.5	Addition and Subraction Using Mixed Numerals
3.6	Multiplication and Division Using Mixed Numerals	3.6	Multiplication and Division Using Mixed Numerals
		(New) 3.7	Order of Operations

Comments: Objective *3.4b* (Writing Mixed Numerals) has been split into two objectives: *3.4b* Writing Mixed Numerals and *3.4c* Finding Mixed Numerals for Quotients. Section *3.7* is new. The concept of averaging fractions included in this new Order of Operations Section.

Chapter 4 - 6th and 7th - Addition and Subtraction: Decimal Noatation

Sec.	Basic Mathematics 6th	Sec.	Basic Mathematics 7th
4.1	Decimal Notation	4.1	Decimal Notation
4.2	Order and Rounding	4.2	Order and Rounding
4.3	Addition and Subtraction with Decimals	4.3	Addition and Subtraction with Decimals
4.4	Solving Problems	4.4	Solving Problems

Comments: In Section *4.4* perimeter is continued (from Section *1.2a)* in Exercise Set 4.4.

Chapter 5 - 6th and 7th - Multiplication and Division: Decimal Notation

Sec.	Basic Mathematics 6th	Sec.	Basic Mathematics 7th
5.1	Multiplication with Decimal Notation	5.1	Multiplication with Decimal Notation
5.2	Division with Decimal Notation	5.2	Division with Decimal Notation
5.3	Converting Fractional Notation to Decimal Notation	5.3	Converting Fractional Notation to Decimal Notation
5.4	Estimating	5.4	Estimating
5.5	Solving Problems	5.5	Solving Problems

Comments: In Section *5.1a* the material is presented in a more logical order. In Section *5.2c* averaging numbers in decimal notation has been added, expanding the drill of order of operations in this objective. In Section *5.3c* the objective was expanded to include order of operations involving fractions and decimals in the same exercise. The number of exercises for this objective has been greatly increased in the exercise set. In Section 5.4a the examples are now in multiple choice form to match the style in the margin exercises.

Chapter 6 - 6th and 7th - Ratio and Proportion

Sec.	Basic Mathematics 6th	Sec.	Basic Mathematics 7th
6.1	Introduction to Ratio and Proportion	6.1	Introduction to Ratio
		6.3	Proportions
6.2	Rates	6.2	Rates
6.3	Proportion Problems	6.4	Proportion Problems
		(New) 6.5	Similar Triangles

Comments: Section 6.1b (Simplifying Notation for Ratios) is new in the Seventh Edition. Sections 6.1b and c of the Sixth Edition are now Sections 6.3a and b of the Seventh Edition. This places the material on proportion after all material on ratios and immediately before solving proportion problems in 6.4. In Section 6.2b the prices have been updated in the unit-pricing objective. Section 6.5 (Similar Triangles) is new in the Seventh Edition. This is an excellent example of proportions and many reviewers requested this topic.

Chapter 7 - 6th and 7th - Percent Notation

Sec.	Basic Mathematics 6th	Sec.	Basic Mathematics 7th
7.1	Percent Notation	7.1	Percent Notation
7.2	Percent Notation and Fractional Notation	7.2	Percent Notation and Fractional Notation
7.3	Solving Percent Problems Using Equations	7.3	Solving Percent Problems Using Equations
7.4	Solving Percent Problems Using Proportions	7.4	Solving Percent Problems Using Proportions
7.5	Applications of Percent	7.5	Applications of Percent
7.6	Consumer Applications: Sales Tax	7.6	Consumer Applications: Sales Tax
7.7	Consumer Applications: Commission and Discount	7.7	Consumer Applications: Commission and Discount
7.8	Consumer Application: Interest	7.8	Consumer Application: Interest

Comments: Sections 7.1b,c and 7.2a have been improved with more in-depth explanation of the rules. In Section 7.2b the margins and exercise set contain more exercises for completing fraction-decimal-% tables.

Chapter 8 - 6th and 7th - Descriptive Statistics

Sec.	Basic Mathematics 6th	Sec.	Basic Mathematics 7th
8.1	Average, Medians and Modes	8.1	Average, Medians and Modes
8.2	Tables, Charts, and Pictographs	8.2	Tables and Pictographs
8.3	Bar Graphs and Line Graphs	8.3	Bar Graphs and Line Graphs
8.4	Circle Graphs	8.4	Circle Graphs

Comments: In Section *8.2* Tables is now used to refer to all tables *and* charts. This should eliminate confusion.

Chapter 9 - 6th and 7th - Geometry and Measures: Length and Area

Sec.	Basic Mathematics 6th	Sec.	Basic Mathematics 7th
9.1	Linear Measures: American Units	9.1	Linear Measures: American Units
9.2	Linear Measures: The Metric System	9.2	Linear Measures: The Metric System
9.3	Perimeter	9.3	Perimeter
9.4	Area	9.4	Area
9.5	Areas of Parallelograms, Triangles, and Trapezoids	9.5	Areas of Parallelograms, Triangles, and Trapezoids
9.6	Circles	9.6	Circles
9.7	Square Roots and the Pythagorean Theorem	9.7	Square Roots and the Pythagorean Theorem

Comments: Chapters 9 and 10: In Section *9.3a*, the definition of a rectangle has been included. Throughout 9 and 10, there is an expanded explanation of when we substitute to convert from one unit to another, and when we multiply by 1 to change units.

Chapter 10 - 6th and 7th - More on Measures

Sec.	Basic Mathematics 6th	Sec.	Basic Mathematics 7th
10.1	Volume and Capacity	10.1	Volume and Capacity
10.2	Volume of Cylinders, Spheres and Cones	10.2	Volume of Cylinders, Spheres and Cones
10.3	Weight, Mass, and Time	10.3	Weight, Mass, and Time
10.4	Temperature	10.4	Temperature
10.5	Converting Units of Area	10.5	Converting Units of Area

Comments: See above.

Chapter 11 - 6th and 7th - The Real-Number System

Sec.	Basic Mathematics 6th	Sec.	Basic Mathematics 7th
11.1	The Real Numbers	11.1	The Real Numbers
11.2	Addition of Real Numbers	11.2	Addition of Real Numbers
11.3	Subtraction of Real Numbers	11.3	Subtraction of Real Numbers
11.4	Multiplication of Real Numbers	11.4	Multiplication of Real Numbers
11.5	Division and Order of Operations	11.5	Division of Real Numbers

Comments: In Section *11.1b* instead of the phrase numbers of arithmetic we now use arithmetic numbers. In Section *11.1d* two pieces of art have been added. (1) number line and (2) family tree chart to explain the subsets of the real numbers. In Section *11.2a* and *11.4a* a new shift arrow is used. Blue to red--positive to negative and Red to blue--negative to positive.

Chapter 12 - 6th and 7th - Algebra: Solving Equations and Problem Solving

Sec.	Basic Mathematics 6th	Sec.	Basic Mathematics 7th
12.1	Introduction To Algebra and Expressions	12.1	Introduction To Algebra
12.2	The Addition Principle	12.2	Solving Equations: The Addition Principle
12.3	The Multiplication Principle	12.3	Solving Equations: The Multiplication Principle
12.4	Using the Principles Together	12.4	Using the Principles Together
12.5	Solving Problems	12.5	Solving Problems

Comments: In Section *12.3a* example 1 has been changed from $3/8 = (-5/4)x$ to $(2/3)x = 18$ to provide an easier lead-in to this Section. Section *12.4c* is new to Section *12.4*. This Section has been expanded to include solving equations by first removing parentheses and collecting like terms. Equations with parentheses were not included in the Sixth Edition.

Developmental Units:
All Low-Stress Sections *A3, S3, M3* have been deleted. Other Sections have been combined.

Sections *A.1* and *A.2* are now in Section *A*.
Sections *S.1* and *S.2* are now in Section *S*.
Sections *M.1* and *M.2* are now in Section *M*.
Sections *D.1* and *D.2* are now in Section *D*.
All objectives in *A.1, A.2, S.1, S.2, M.1, M.2, D.1,* and *D.2* remain the same in Sections *A, S, M,* and *D*.

Basic Mathematics Video Index

The following is a list of text exercises which appear on the Math Makes a
Difference videos:

Text/Video Section	Exercise Numbers
Section 1.1	3, 21, 37, 43, 45
Section 1.2	7, 9, 37, 43
Section 1.3	39
Section 1.4	23, 29
Section 1.5	21
Section 1.6	11
Section 1.7	1, 5, 7, 41, 45, 55, 57
Section 1.8	23
Section 1.9	5, 7, 9, 11, 35, 51
Section 2.1	13, 47, 59, 65
Section 2.2	1, 3, 5, 7, 9, 11, 15
Section 2.3	21
Section 2.4	1, 5, 27, 37
Section 2.5	3, 21, 33, 37, 39
Section 2.6	1, 7, 9, 19, 35, 45, 53
Section 2.7	9, 13, 19, 21, 33, 35, 45
Section 3.1	1, 3, 7, 9, 19, 23, 29, 39
Section 3.2	3, 5, 11, 15, 29, 33, 40
Section 3.3	9, 29, 43, 53
Section 3.4	7, 15, 23, 41, 43, 51
Section 3.5	1, 9, 15, 17, 19, 23, 35
Section 3.6	3, 7, 20, 26, 37, 43
Section 3.7	7, 11, 21, 27
Section 4.1	9, 19, 23, 33, 37
Section 4.2	3, 7, 11, 37, 53
Section 4.3	9, 11, 15, 31, 61, 63, 67
Section 4.4	6, 16, 23, 24, 25, 26
Section 5.1	3, 5, 7, 23, 33, 57
Section 5.2	33, 35, 59, 69
Section 5.3	1, 3, 15, 23, 32
Section 5.4	1, 5, 23
Section 5.5	7, 40
Section 6.1	3, 7, 13, 22, 23, 31, 35
Section 6.2	1, 3, 7, 9, 21
Section 6.3	1, 5, 13, 19, 21, 29, 35, 37
Section 6.4	3, 15, 17
Section 6.5	9
Section 7.1	1, 23, 47, 51
Section 7.2	11, 15, 19, 32
Section 7.3	3, 5, 23, 25, 31
Section 7.4	3, 5, 13, 17

Section 7.5	3, 15, 17, 19
Section 7.6	1, 5
Section 7.7	9
Section 7.8	none
Section 8.1	1, 7, 20
Section 8.2	none
Section 8.3	11, 13, 15, 25
Section 8.4	3
Section 9.1	8, 11, 17, 21, 27, 35
Section 9.2	7, 21, 31, 35, 41
Section 9.3	3, 5, 15
Section 9.4	13, 21
Section 9.5	1, 9, 19
Section 9.6	25, 31
Section 9.7	1, 3, 17, 19, 25, 29, 33, 45
Section 10.1	1, 19, 25
Section 10.2	15, 21, 23
Section 10.3	5, 51
Section 10.4	19, 21
Section 10.5	1, 15, 19, 27
Section 11.1	5, 15, 21, 22, 31, 37, 43, 45, 51
Section 11.2	11, 15, 31, 39, 47, 49, 63, 65, 67
Section 11.3	1, 3, 7, 9, 31, 39, 57, 67, 71
Section 11.4	7, 13, 22, 27
Section 11.5	9, 11, 15, 17, 21, 23, 25, 37, 75
Section 12.1	3, 9, 13, 23, 29, 33, 37, 43, 45, 47, 49, 59
Section 12.2	5, 15, 17, 19, 21
Section 12.3	1, 7, 9, 13, 23, 25
Section 12.4	7, 11, 17, 35, 41, 47, 53
Section 12.5	1, 3, 6, 15, 17, 24, 31, 37

Tape 20

INDEX

Bittinger/Keedy, 7/e

Audiotape Series

<u>BASIC</u> <u>MATHEMATICS</u>

Tape		Side A	Tape		Side B
Tape 1	Sections	1.1-1.2	Tape 1	Sections	1.3-1.4
Tape 2	Sections	1.5-1.6	Tape 2	Sections	1.7-1.8
Tape 3	Sections	1.9-2.1	Tape 3	Sections	2.2-2.3
Tape 4	Sections	2.4-2.5	Tape 4	Sections	2.6-2.7
Tape 5	Sections	3.1-3.2	Tape 5	Sections	3.3-3.4
Tape 6	Sections	3.5-3.6	Tape 6	Sections	3.7-4.1
Tape 7	Sections	4.2-4.3	Tape 7	Sections	4.4-5.1
Tape 8	Sections	5.2-5.3	Tape 8	Sections	5.4-5.5
Tape 9	Sections	6.1-6.2	Tape 9	Sections	6.3-6.4
Tape 10	Sections	6.5-7.1	Tape 10	Sections	7.2-7.3
Tape 11	Sections	7.4-7.5	Tape 11	Sections	7.6-7.7
Tape 12	Sections	7.8-8.1	Tape 12	Sections	8.2-8.3
Tape 13	Sections	8.4-9.1	Tape 13	Sections	9.2-9.3
Tape 14	Sections	9.4-9.5	Tape 14	Sections	9.6-9.7
Tape 15	Sections	10.1-10.2	Tape 15	Sections	10.3-10.4
Tape 16	Sections	10.5-11.1	Tape 16	Sections	11.2-11.3
Tape 17	Sections	11.4-11.5	Tape 17	Sections	12.1-12.2
Tape 18	Sections	12.3-12.4	Tape 18	Sections	12.5
Tape 19	Appendixes	A-S	Tape 19	Appendixes	M-D

InterAct Math Tutorial Software

Following are section/topic cross references for the **InterAct Math Tutorial Software** that was developed to accompany *Basic Mathematics, Seventh Edition*. **InterAct Math Tutorial Software** includes exercises that are linked one-to-one with the odd-numbered exercises in the textbook and require the same computational and problem-solving skills as their companion exercises in the text.

Each exercise has an example and an interactive guided solution that are designed to involve students in the solution process and help them identify precisely where they are having trouble. In addition, the software recognizes common student errors and provides students with appropriate customized feedback.

InterAct Math Tutorial Software has sophisticated answer recognition capabilities that enable it to recognize appropriate forms of the same answer for any kind of input. It also tracks student activity and scores for each section, which can be printed out.

Icons at the beginning of each text section reference the appropriate disk number. Available for both DOS-based and Macintosh computers, the software is free to adopters.

Hardware Requirements:

InterAct Math Tutorial Software, DOS Version
XT, AT, 286 or higher IBM PC or IBM PC compatible with at least one floppy disk drive*. DOS 3.3 or higher 640K RAM, 530K RAM free, 550K recommended, CGA, EGA, VGA, SVGA graphics adapter

*Also can be installed on a network or hard drive.

InterAct Math Tutorial Software, Macintosh Version
Macintosh Plus or newer computer 1 Megabyte of RAM. System 6.0.3 or higher

To receive demonstration disks for **InterAct Math Tutorial Software**, please contact your Addison-Wesley sales representative, or one of the regional sales offices.

Disk 7
    7.1   Percent Notation
    7.2   Percent Notation and Fractional Notation
    7.3   Solving Percent Problems Using Equations
    7.4   Solving Percent Problems Using Proportions
    7.5   Applications of Percent
    7.6   Consumer Applications: Sales and Tax
    7.7   Consumer Applications: Commission and Discount
    7.8   Consumer Applications: Interest

Disk 8
    8.1   Averages, Medians, and Modes
    8.2   Tables and Pictograph
    8.3   Bar Graphs and Line Graphs
    8.4   Circle Graphs

Disk 9
    9.1   Linear Measures: American Units
    9.2   Linear Measures: The Metric System
    9.3   Perimeter
    9.4   Area
    9.5   Areas of Parallelograms, Triangles and Trapezoids
    9.6   Circles
    9.7   Square Roots and The Pythagorean Theorem

Disk 10
    10.1 Volume and Capacity
    10.2 Volume of Cylinders, Spheres, and Cones
    10.3 Weight, Mass, and Time
    10.4 Temperature
    10.5 Converting Units of Area

Disk 11A
    11.1 The Real Numbers
    11.2 Addition of Real Numbers
    11.3 Subtraction of Real Numbers
    11.4 Multiplication of Real Numbers

Disk 11B
    11.5 Division of Real Numbers

Disk 12
    12.1 Introduction to Algebra
    12.2 Solving Equations: The Addition Principle
    12.3 Solving Equations: The Multiplication Principle
    12.4 Using the Principles Together
    12.5 Solving Problems

InterAct Math
5.25 MS-DOS
Basic Mathematics 7/e

Disk 5B
- 5.3   Converting from Fractional Notation to Decimal Notation
- 5.4   Estimating
- 5.5   Solving Problems

Disk 6
- 6.1   Introduction to Ratios
- 6.2   Rates
- 6.3   Proportions
- 6.4   Proportion Problems
- 6.5   Similar Triangles

Disk 7A
- 7.1   Percent Notation
- 7.2   Percent Notation and Fractional Notation
- 7.3   Solving Percent Problems Using Equations

Disk 7B
- 7.4   Solving Percent Problems Using Proportions
- 7.5   Applications of Percent
- 7.6   Consumer Applications: Sales and Tax
- 7.7   Consumer Applications: Commission and Discount
- 7.8   Consumer Applications: Interest

Disk 8
- 8.1   Averages, Medians, and Modes
- 8.2   Tables and Pictograph
- 8.3   Bar Graphs and Line Graphs
- 8.4   Circle Graphs

Disk 9A
- 9.1   Linear Measures: American Units
- 9.2   Linear Measures: The Metric System
- 9.3   Perimeter
- 9.4   Area
- 9.5   Areas of Parallelograms, Triangles and Trapezoids

Disk 9B
- 9.6   Circles
- 9.7   Square Roots and The Pythagorean Theorem

Disk 10A
- 10.1 Volume and Capacity
- 10.2 Volume of Cylinders, Spheres, and Cones
- 10.3 Weight, Mass, and Time
- 10.4 Temperature

Disk 10B
- 10.5 Converting Units of Area

InterAct Math
5.25 MS-DOS
Basic Mathematics 7/e

InterAct Math
Macintosh
Basic Mathematics 7/e

Disk 1A
- 1.1 Standard Notation
- 1.2 Addition
- 1.3 Subtraction
- 1.4 Rounding and Estimating; Order

Disk 1B
- 1.5 Multiplication
- 1.6 Division
- 1.7 Solving Equations
- 1.8 Solving Problems
- 1.9 Exponential Notation and Order of Operations

Disk 2
- 2.1 Factorizations
- 2.2 Divisibility
- 2.3 Fractions
- 2.4 Multiplication
- 2.5 Simplifying
- 2.6 Multiplying and Simplifying
- 2.7 Reciprocals and Division

Disk 3A
- 3.1 Least Common Multiples
- 3.2 Addition
- 3.3 Subtraction and Order
- 3.4 Mixed Numerals
- 3.5 Addition and Subtraction Using Mixed Numerals
- 3.6 Multiplication and Division Using Mixed Numerals

Disk 3B
- 3.7 Order of Operations

Disk 4
- 4.1 Decimal Notation
- 4.2 Ordering and Rounding
- 4.3 Addition and Subtraction with Decimals
- 4.4 Solving Problems

Disk 5
- 5.1 Multiplication with Decimal Notation
- 5.2 Division with Decimal Notation
- 5.3 Converting from Fractional Notation to Decimal Notation
- 5.4 Estimating
- 5.5 Solving Problems

Disk 6
- 6.1 Introduction to Ratios
- 6.2 Rates
- 6.3 Proportions
- 6.4 Proportion Problems
- 6.5 Similar Triangles

Disk 7
  7.1 Percent Notation
  7.2 Percent Notation and Fractional Notation
  7.3 Solving Percent Problems Using Equations
  7.4 Solving Percent Problems Using Proportions
  7.5 Applications of Percent
  7.6 Consumer Applications: Sales and Tax
  7.7 Consumer Applications: Commission and Discount
  7.8 Consumer Applications: Interest

Disk 8
  8.1 Averages, Medians, and Modes
  8.2 Tables and Pictograph
  8.3 Bar Graphs and Line Graphs
  8.4 Circle Graphs

Disk 9
  9.1 Linear Measures: American Units
  9.2 Linear Measures: The Metric System
  9.3 Perimeter
  9.4 Area
  9.5 Areas of Parallelograms, Triangles and Trapezoids
  9.6 Circles
  9.7 Square Roots and The Pythagorean Theorem

Disk 10
  10.1 Volume and Capacity
  10.2 Volume of Cylinders, Spheres, and Cones
  10.3 Weight, Mass, and Time
  10.4 Temperature
  10.5 Converting Units of Area

Disk 11A
  11.1 The Real Numbers
  11.2 Addition of Real Numbers
  11.3 Subtraction of Real Numbers
  11.4 Multiplication of Real Numbers

Disk 11B
  11.5 Division of Real Numbers

Disk 12
  12.1 Introduction to Algebra
  12.2 Solving Equations: The Addition Principle
  12.3 Solving Equations: The Multiplication Principle
  12.4 Using the Principles Together
  12.5 Solving Problems

Following is a cross-reference keying the topics from *Basic Mathematics, Seventh Edition* to the contents of MODUMATH. MODUMATH is interactive laserdisc courseware that is available from VTAE (One Foundation Circle, Waunakee, WI. 53597-8914. Phone 800-821-6313).

ModuMath 1.1 - Naming Whole Numbers

When you have completed this lesson you will be able to:

1) Read decimal numbers.
2) Change whole numbers into word form.
3) Change whole numbers stated in words into decimal numeral form.
4) Write decimal numerals in expanded notation.
5) State the number represented by each digit in a decimal.

Section One - Digits
Section Two - Place Value and Face Value
Section Three - Expanded Notation
Section Four - Reading Large Numbers

ModuMath 1.2 - The Number Line

When you have completed this lesson, you will be able to:

1) Construct a number line.
2) Graph whole numbers on a number line.
3) Find the coordinate of a point on a number line.
4) Decide which of two points represents the smaller (or larger) whole number.
5) Use the < or > symbols to compare whole numbers.

Section One - Drawing the Number Line and Graphing Numbers
Section Two - Graphs and Coordinates
Section Three - Greater or Less Than

ModuMath 1.3 - Addition of Whole Numbers, Part I

When you have completed this lesson you will be able to:

1) State the 100 addition facts from memory.
2) Interpret the addition of whole numbers on the number line.
3) Recognize and use the identity for additon.
4) Recognize and solve word problems involving addition of small whole numbers.

In order to successfully complete this lesson, you should be familiar with:

1) The Number Line. (ModuMath 1.2)

Section One - The Basics of Addition
Section Two - Practical Addition

ModuMath 1.4 - Addition of Whole Numbers, Part II

When you have completed this lesson you will be able to:

1) Find the sum of two or more whole numbers.

In order to successfully complete this lesson, you should be familiar with:

1) Expanded Notation (ModuMath 1.1)
2) Addition of Whole Numbers, Part I (ModuMath 1.3)

Section One - Adding Large Whole Numbers
Section Two - Practical Problems

ModuMath 1.5 - Subtracting Whole Numbers

When you have completed this lesson you will be able to:

1) State from memory the hundred subtraction facts.
2) Subtract whole numbers, providing the first is greater than or equal to the second.
3) Check a subtraction problem by addition.
4) Solve word problems involving subtraction.

In order to successfully complete this lesson, you should be familiar with:

1) Expanded Notation (ModuMath 1.1)
2) The Number Line (ModuMath 1.2)
3) Additon of Whole Numbers (ModuMath 1.3, 1.4)

Section One - Subtraction of a One Digit Number
Section Two - Subtracting Larger Numbers
Section Three - Exchanging (Borrowing) and Checking By Arithmetic
Section Four - Practical Problems

ModuMath 1.6 - Multiplying Whole Numbers, Part I

When you have completed this lesson you will be able to:

1) Use multiplication as a shortcut for addition.
2) Interpret multiplication on the number line.
3) Show multiplication facts as arrays.
4) Use multiplication to find areas.
5) Use multiplication to figure out combinations.
6) Multiply by zero.
7) Recognize and use the identity for multiplication.

In order to successfully complete this lesson, you should be familiar with:

1) The Number Line (ModuMath 1.2)
2) Addition of Whole Numbers (ModuMath 1.3)

ModuMath 1.7 - Multiplying Whole Numbers, Part II

When you have completed this lesson you will be able to:

1) State the 100 multiplications facts from memory.
2) Solve word problems involving multiplication of small whole numbers.

In order to successfully complete this lesson, you should be familiar with:

1) The Number Line (ModuMath 1.2)
2) Addition of Whole Numbers (ModuMath 1.3, 1.4)
3) Multiplying Whole Numbers, Part I (ModuMath 1.6)

Section One - Introducing Multiplication
Section Two - Practical Multiplication Problems

ModuMath 1.8 - Multiplying Whole Numbers, Part III

When you have completed this lesson you will be able to:

1) Multiply any two whole numbers.
2) Solve word problems involving multiplication.

In order to successfully complete this lesson, you should be familiar with:

1) Multiplication of One Digit Numbers (ModuMath 1.7)
2) Splitting Arrays (ModuMath 1.6, 1.7)
3) Expanded Notation (ModuMath 1.1)

Section One - Some Multiplication Tricks
Section Two - Multiplying Large Whole Numbers

ModuMath 1.9 - Exponents

When you have completed this lesson you will be able to:

1) Read expressions involving exponents.
2) Multiply two or more expressions with the same base by adding exponents.
3) Recognize and work with expressions having exponents equal to zero.
4) Write whole numbers in exponential notation.

In order to successfully complete this lesson, you should be familiar with:

1) Multiplying Whole Numbers (ModuMath 1.7, 1.8)
2) Expanded Notation (ModuMath 1.1)

Section One - Factor, Power, Base, and Exponent
Section Two - Square and Square Root
Section Three - Adding Exponents
Section Four - Exponents of 0
Section Five - Expanded Notation

ModuMath 1.10 - Dividing Whole Numbers, Part I

When you have completed this lesson you will be able to:

1) State the ninety basic division facts from memory.
2) Solve simple word problems involving division.
3) Use multiplication to check division.

In order to successfully complete this lesson, you should be familiar with:

1) Subtract Whole Numbers. (ModuMath 1.5)
2) Multiply Whole Numbers. (ModuMath 1.7)

Section One - Introducing Division
Section Two - The Rules of Division
Section Three - Remainders and Some Practical Problems

ModuMath 1.11 - Dividing Whole Numbers, Part II

When you have completed this lesson you will be able to:

1) Divide any whole number by any other.
2) Solve word problems involving division.
3) Use multiplication to check division.

In order to successfully complete this lesson, you should be familiar with:

1) Dividing Whole Numbers, Part I (ModuMath 1.10)
2) Multiplying Whole Numbers (ModuMath 1.8)
3) Subtracting Whole Numbers (ModuMath 1.5)
4) Place Value (ModuMath 1.1)

Section One - Review and New Ideas About Division
Section Two - Using Partial Quotients
Section Three - Other Methods of Division

ModuMath 1.12 - Word Problems

When you have completed this lesson you will be able to:

1) Express word problems as number sentences.
2) Solve word problems involving addition, subtraction, multiplication, and
   division of whole numbers.

In order to successfully complete this lesson, you should be familiar with:

1) Adding Whole Numbers (ModuMath 1.3, 1.4)
2) Subtracting Whole Numbers (ModuMath 1.5)
3) Multiplying Whole Numbers (ModuMath 1.7, 1.8)
4) Dividing Whole Numbers (ModuMath 1.10, 1.11)

Section One - Using Arithmetic to Solve Practical Problems
Section Two - More Word Problems

ModuMath 1.13 - Solving Equations

When you have completed this lesson you will be able to:

1) Solve simple equations.
2) Solve word problems involving simple questions.

In order to successfully complete this lesson, you should be familiar with:

1) Add whole numbers (ModuMath 1.4)
2) Subtract whole numbers (ModuMath 1.5)
3) Multiply whole numbers (ModuMath 1.8)
4) Divide whole numbers (Modumath 1.10)
5) Deal with word problems (ModuMath 1.12)

Section One - Working With Equations
Section Two - Word Problems and Equations

ModuMath 1.14 - Prime Numbers

When you have completed this lesson you will be able to:

1) Determine whether a given whole number greater than 1 is prime or composite.
2) Use the Sieve of Eratosthenes.
3) Determine the prime factorization of a given whole number.

In order to successfully complete this lesson, you should be familiar with:

1) Multiplication of Whole Numbers (ModuMath 1.6, 1.7, 1.8, 1.9)
2) Division of Whole Numbers (ModuMath 1.10, 1.11)

Section One - Determining Prime or Composite Numbers
Section Two - Sieve of Eratosthenes
Section Three - Prime Factorization

ModuMath 2.1 - Introducing Fractions

When you have completed this lesson you will be able to:

1) Use fractions in comparing two quantities.

2) Use fractions in expressing parts of a whole.
3) Use fractions in expressing division.
4) Use fractions in measurement.
5) Locate fractions on the number line.
6) Recognize and use fractions with a denominator of 1.
7) Divide whole numbers, expressing the quotient as a fraction or as a whole number plus a fraction.

In order to successfully complete this lesson, you should be familiar with:

1) The Number Line (ModuMath 1.2)
2) Division of Whole Numbers (ModuMath 1.10 and 1.11)

Section One - Fractions in Comparing Two Quantites
Section Two - Parts of the Whole
Section Three - Division and Fractions
Section Four - Locating Numbers on the Number Line
Section Five - Measurement

ModuMath 2.2 - Renaming Fractions

When you have completed this lesson you will be able to:

1) Express fractions in higher or lower terms.
2) Reduce fractions to lowest terms
3) Recognize and use fractions equal to 1.
4) Recognize and use fractions equal to 0.

In order to successfully complete this lesson, you should be familiar with:

1) Prime Numbers (ModuMath 1.14 - VERY IMPORTANT)
2) The Number Line (ModuMath 1.2)
3) Multiplication of Whole Numbers (ModuMath 1.7, 1.8)
4) Division of Whole Numbers (ModuMath 1.10, 1.11)
5) Introducing Fractions (ModuMath 2.1)

Section One - Fractions That Name the Same Point
Section Two - Expressing Fractions in Higher or Lower Terms
Section Three - Fractions That Equal 1 or 0.

ModuMath 2.3 - Adding Fractions with the Same Denominator

When you have completed this lesson you will be able to:

1) Add two or more fractions with the same denominator.
2) Change a mixed number to an improper fraction.
3) Change an improper fraction to a mixed number.
4) Add any combination of mixed numbers, fractions, and whole numbers, provided all the denominators are the same.

In order to successfully complete this lesson, you should be familiar with:

1) Addition of Whole Numbers (ModuMath 1.3, 1.4)
2) Renaming Fractions (ModuMath 2.2)
3) Multiplication of Whole Numbers (ModuMath 1.6, 1.7)
4) Division of Whole Numbers (ModuMath 1.12, 1.13)

Section One - Adding Fractions Whose Sum is a Proper Fraction
Section Two - Improper and Proper Fractions
Section Three - Changing Mixed Numbers to Improper Fractions

Section Four - Changing Improper Fractions to Mixed Numbers
Section Five - Adding Fractions Whose Sums Are Improper Fractions
Section Six - Adding Mixed Numbers

ModuMath 2.4 - Adding Fractions with Different Denominators

When you have completed this lesson you will be able to:

1) Find a common denominator of two or more fractions.
2) Find the least common denominator of two or more fractions.
3) Add two or more fractions with different denominators.

In order to successfully complete this lesson, you should be familiar with:

1) Addition of Whole Numbers (ModuMath 1.3, 1.4)
2) Multiplication of Whole Numbers (ModuMath 1.6, 1.7)
3) Division of Whole Numbers (ModuMath 1.12, 1.13)
4) Prime Numbers (ModuMath 1.14)
5) Renaming Fractions (ModuMath 2.2)
6) Adding Fractions with the Same Denominator (ModuMath 2.3)

Section One - Common Denominators by Comparing Multiples
Section Two - Shortcut to Common Denominators
Section Three - Lowest Common Denominator Using Primes

ModuMath 2.5 - Subtracting Fractions

When you have completed this lesson you will be able to:

1) Indicate the larger of two fractions.
2) Subtract fractions.
3) Subtract mixed numbers.
4) Solve word problems involving subtraction of fractions.

In order to successfully complete this lesson, you should be familiar with:

1) The Number Line (<ModuMath 1.2)
2) Subtracting Whole Numbers (ModuMath 1.5)
3) Renaming Fractions (ModuMath 2.2)
4) Changing Improper Fractions to Mixed Numbers (ModuMath 2.3)
5) Common Denominators (ModuMath 2.4)

Section One - Indicating the Larger of Two Fractions
Section Two - Subtracting Fractions
Section Three - Subtracting Mixed Numbers
Section Four - Word Problems

ModuMath 2.6 - Multiplying Fractions

When you have completed this lesson you will be able to:

1) Multiply a fraction or mixed number and a whole number.
2) Multiply any combination of fractions and mixed numbers.
3) Solve word problems involving multiplication of fractions.

In order to successfully complete this lesson, you should be familiar with:

1) Multiplying Whole Numbers (<ModuMath 1.7, 1.8)
2) Renaming Fractions (ModuMath 2.2)
3) Addition of Fractions (ModuMath 2.3)

ModuMath 2.7 - Dividing Fractions

When you have completed this lesson you will be able to:

1) Divide a fraction by a mixed number, a whole number, or another fraction.
2) Divide a mixed number by a fraction, a whole number, or another mixed number.
3) Divide a whole number by a fraction or a mixed number.
4) Solve word problems involving division of fractions.

In order to successfully complete this lesson, you should be familiar with:

1) Multiplying Whole Numbers (<ModuMath 1.8)
2) Dividing Whole Numbers (ModuMath 1.10)
3) Multiplying Fractions (ModuMath 2.6)

ModuMath 2.8 - Ratio and Proportion

When you have completed this lesson you will be able to:

1) Determine whether two ratios are equal.
2) Solve word problems involving proportions.

In order to successfully complete this lesson, you should be familiar with:

1) Multiplying Whole Numbers (<ModuMath 1.8)
2) Word Problems (ModuMath 1.12)
3) Solving Equations (ModuMath 1.13)
4) Understanding Fractions (ModuMath 2.1)
5) Renaming Fractions (ModuMath 2.2)

ModuMath 3.1 - Understanding Decimal Fractions

When you have completed this lesson you will be able to:

1) Write decimal fractions in exponential notation.
2) Read decimal fractions.
3) Change decimals to fractions.
4) Change a fraction to a decimal, provided the denominator is a power of ten.
5) Recognize the meaning of zeros to the right of a decimal.

In order to successfully complete this lesson, you should be familiar with:

1) Multiplying Whole Numbers (ModuMath 1.7)
2) Exponents (ModuMath 1.9)
3) Uses of Fractions (ModuMath 2.1)

Section One - Introducing Decimal Fractions
Section Two - Using Decimal Fractions

ModuMath 3.2 - Adding and Subtracting Decimal Fractions

When you have completed this lesson you will be able to:

1) Determine which is the larger or smaller of two decimal fractions.
2) Add two or more decimal fractions.
3) Subtract decimal fractions.
4) Solve word problems involving addition and subtraction of decimal fractions.

In order to successfully complete this lesson, you should be familiar with:

1) Addition of Whole Numbers (ModuMath 1.4)
2) Subtraction of Whole Numbers (ModuMath 1.5)
3) Addition of Fractions (ModuMath 2.3)
4) Decimal Fractions (ModuMath 3.1)

Section One - Adding Decimal Fractions
Section Two - Subtracting Decimal Fractions

ModuMath 3.3 - Rounding Numbers

When you have completed this lesson you will be able to:

1) Round a number to a given degree of accuracy.

In order to successfully complete this lesson, you should be familiar with:

1) Understanding Decimal Fractions (ModuMath 3.1)
2) Naming Whole Numbers (ModuMath 1.1)

Section One - How Numbers are Rounded
Section Two - Exercises in Rounding Numbers

ModuMath 3.4 - Multiplying Decimals

When you have completed this lesson you will be able to:

1) Multiply decimal fractions.
2) Solve word problems involving multiplication of decimals.

In order to successfully complete this lesson, you should be familiar with:

1) Multiplying Whole Numbers (ModuMath 1.8)
2) Exponents (ModuMath 1.9)
3) Understanding Decimal Fractions (ModuMath 3.1)

Section One - How to Multiply Decimals
Section Two - Why It's Done This Way

ModuMath 3.5 - Dividing Decimal Fractions

When you have completed this lesson you will be able to:

1) Divide a decimal by a whole number.
2) Divide a whole number by a decimal.
3) Divide a decimal by another decimal.
4) Divide accurately to a given number of decimal places.
5) Solve word problems involving division of decimals.

In order to successfully complete this lesson, you should be familiar with:

1) Multiplication of Decimals (ModuMath 3.4)
2) Division of Whole Numbers (ModuMath 1.11)
3) Fractions (ModuMath 2.1, 2.2)

Section One - Introduction
Section Two - The Four Possibilities

ModuMath 3.6 - Changing Fractions to Decimals

When you have completed this lesson you will be able to:

1) Change a fraction or mixed number to a decimal.
2) Change a decimal to a fraction or mixed number

In order to successfully complete this lesson, you should be familiar with:

1) Division of Decimals (ModuMath 3.5)
2) Rounding Decimals (ModuMath 3.3)
3) Fractions (ModuMath 2.1)

ModuMath 3.7 - Square Roots

When you have completed this lesson you will be able to:

1) Use a table to find square roots of whole numbers from 1 to 100.
2) Recognize and use the fact that if a and b are any numbers, then the square
   root of a times b is the same as the square root of a times the square root
   of b.

In order to successfully complete this lesson, you should be familiar with:

1) Multiplication of Whole Numbers (ModuMath 1.8)
2) Exponents (ModuMath 1.9)
3) Multiplication of Decimals (ModuMath 3.4)

Section One - Square Root Fundamentals
Section Two - Practical Square Root Problems

ModuMath 4.1 - The Meaning of Percentage

When you have completed this lesson you will be able to:

1) Change fractions and decimals to percents.
2) Change percents to fractions and decimals.

In order to successfully complete this lesson, you should be familiar with:

1) Change fractions to decimals. (ModuMath 3.6)
2) Understand ratios and proportions. (ModuMath 2.8)

Section One - The Meaning of Percentage
Section Two - Working with Percent

ModuMath 4.2 - Percent in Word Problems

When you have completed this lesson you will be able to:

1) Solve word problems involving percents by using proportions.

In order to successfully complete this lesson, you should be familiar with:

1) Percent (ModuMath 4.1)
2) Ratio and Proportion (ModuMath 2.8)
3) Word Problems (ModuMath 1.12)

Section One - Using Ratio and Proportion with Percent
Section Two - Profit, Discount and Commission

ModuMath 4.3 - More Problems in Percent

When you have completed this lesson you will be able to:

1) Solve more complicated word problems involving percents.

In order to successfully complete this lesson, you should be familiar with:

1) Percent in Word Problems (ModuMath 4.2)
2) Percent (ModuMath 4.1)
3) Word Problems (ModuMath 1.12)

Section One - Practice Problems
Section Two - Population and Interest

ModuMath 5.1 - Signed Numbers

When you have completed this lesson you will be able to:

1) Graph positive and negative numbers.
2) Decide which is the larger (or smaller) of two given numbers.
3) Find the absolute value of any given number.

In order to successfully complete this lesson, you should be familiar with:

1) The Number Line (ModuMath 1.2)

Section One - Graphing Positive and Negative Numbers
Section Two - Absolute Value
Section Three - Greater or Less Than, Using Absolute Values

ModuMath 5.2 - Adding Signed Numbers

When you have completed this lesson you will be able to:

1) Add two or more signed numbers.
2) Solve word problems involving addition of signed numbers.

In order to successfully complete this lesson, you should be familiar with:

1) Addition of Whole Numbers (ModuMath 1.4)
2) Subtraction of Whole Numbers (ModuMath 1.5)
3) Signed Numbers on the Number Line (ModuMath 5.1)

Section One - Adding Numbers with the Same Signs
Section Two - Adding Numbers with Unlike Signs
Section Three - Adding More Than Two Numbers
Section Four - Practical Problems

ModuMath 5.3 - Subtracting Signed Numbers

When you have completed this lesson you will be able to:

1) Subtract any two signed numbers.

In order to successfully complete this lesson, you should be familiar with:

1) Subtacting Whole Numbers (ModuMath 1.5)
2) Signed Numbers on the Number Line (ModuMath 5.1)
3) Adding Signed Numbers (ModuMath 5.2)

Section One - Subtracting Signed Numbers, Using the Number Line
Section Two - Rules for Subtraction

ModuMath 5.4 - Multiplying Signed Numbers

When you have completed this lesson you will be able to:

1) Multiply two or more signed numbers.

In order to successfully complete this lesson, you should be familiar with:

1) Multiplying Whole Numbers (ModuMath 1.6, 1.7, 1.8)
2) Adding Signed Numbers (ModuMath 5.2)

Section One - Multiply Numbers with Different Signs
Section Two - Multiplying Negative Times a Negative
Section Three - Multiplying Three or More Signed Numbers

ModuMath 5.5 - Dividing Signed Numbers

When you have completed this lesson you will be able to:

1) Divide a given signed number by another signed number.

In order to successfully complete this lesson, you should be familiar with:

1) Dividing Whole Numbers (ModuMath 1.11)
2) Multiplying Signed Numbers (ModuMath 5.4)

Section One - Rules for Dividing Signed Numbers
Section Two - Reasons for the Rules for Division

ModuMath 5.6 - Signed Fractions

When you have completed this lesson you will be able to:

1) Recognize when signed fractions are equal.
2) Add signed fractions.
3) Subtract signed fractions.
4) Multiply signed fractions.
5) Divide signed fractions.
6) Solve equations involving signed fractions.

In order to successfully complete this lesson, you should be familiar with:

1) Solving Equations (ModuMath 1.13)
2) Fractions (ModuMath 2.1 through 2.7)
3) Absolute Value (ModuMath 5.1)
4) Adding, subtracting, multiplying, and dividing signed numbers (ModuMath 5.2, 5.3, 5.4, and 5.5)

Section One - Meaning of Negative Fractions
Section Two - Arithmetic with Signed Numbers
Section Three - Solving Equations

ModuMath 5.7 - Negative Exponents

When you have completed this lesson you will be able to:

1) Recognize that an expression such as 6^{-4} means $\dfrac{1}{6^4}$.

2) Divide two exponents involving exponents when the bases are the same.
3) Write numbers in exponential notation using positive and negative exponents.

In order to successfully complete this lesson, you should be familiar with:

1) Exponents (ModuMath 1.9)
2) Exponential Notation (as discussed in ModuMath 1.1, 3.1)

Section One - Negative Exponents
Section Two - Dividing Using Negative Exponents
Section Three - Expanded Notation

Algebra 1 - Getting Acquainted With Algebra

When you have completed this lesson you will be able to:

1) Translate words into the language of algebra.
2) Evaluate algebraic expressions.
3) Evaluate formulas and other expressions.

In order to successfully complete this lesson, you should be familiar with:

1) Addition, subtraction, multiplication and division.
2) Fractions and signed numbers.
3) Solving word problems.

Section One - What is Algebra?
Section Two - Algebraic Expressions and Equations
Section Three - Formulas

Algebra 2 - Order of Operations

When you have completed this lesson you will be able to:

1) Use the standard order of operations to simplify and evaluate numerical or algebraic expressions.
2) Simplify and evaluate numerical and algebraic expressions that contain parentheses.

In order to successfully complete this lesson, you should be familiar with:

1) Know how to add, subtract, multiply and divide signed numbers.
2) Know what is meant by an algebraic expression.
3) Be familiar with the idea of exponents, and square and cube roots.

Section One - Order of Operations
Section Two - Use of Parentheses

Algebra 3 - Adding and Subtracting Algebraic Expressions

When you have completed this lesson you will be able to:

1) Identify the terms in algebraic expressions.
2) Identify like terms in algebraic expressions.
3) Identify the special types of polynomials called monomials, binomials, and trinomials.
4) Add and subtract any type of polynomials.

In order to successfully complete this lesson, you should be familiar with:

1) Adding and subtracting signed numbers.

Section One - Like terms
Section Two - Adding Algebraic Expressions
Section Three - Subtracting Algebraic Expressions
Section Four - Polynomials

Algebra 4 - Multiplying Polynomials

When you have completed this lesson you will be able to:

1) Multiply any two polynomials.
2) Apply the FOIL method as a shortcut in multiplying two binomials.
3) Use formulas to obtain special products of the form

$$(a + b)^2, \ (a - b)^2, \text{ and } (a + b)(a - b).$$

In order to successfully complete this lesson, you should be familiar with:

1) How to add, subtract, and multiply signed numbers.
2) The laws for multiplying expressions with exponents.
3) The definition of a monomial, binomial, trinomial, and polynomial.
4) How to simplify polynomials by combining like terms.

Section One - Multiplying Signed Numbers and Variables
Section Two - Multiplying By Monomials
Section Three - Multiplying By Binomials

Algebra 5 - Laws of Algebra

When you have completed this lesson you will be able to:

1) Recognize five basic laws of algebra.
2) Apply these laws to certain algebraic operations.

In order to successfully complete this lesson, you should be familiar with:

1) Adding, subtracting, and multiply signed numbers.
2) Adding, subtracting, and multiply polynomials.

Section One - The Commutative Laws for Addition and Multiplication
Section Two - The Associative Laws for Addition and Multiplication
Section Three - The Distributive Law

Algebra 6 - Solving Equations I

When you have completed this lesson you will be able to:

1) Recognize the identities for addition and multiplication.
2) Recognize the inverses for addition and multiplication.
3) Use the identities and inverses in solving equations.

In order to successfully complete this lesson, you should be familiar with:

1) Operations with signed numbers.
2) The associative and commutative laws.

Section One - Identities
Section Two - Tactics for Solving Equations
Section Three - Inverses
Section Four - Equation Identities and Impossibilities

Algebra 7 - Solving Equations II

When you have completed this lesson you will be able to:

1) Solve multiple step first degree equations.
2) Use some shortcuts to solve equations.
3) Understand and work with literal equations.

In order to successfully complete this lesson, you should be familiar with:

1) Solving equations that require only one step.

Section One - Strategies for Solving Equations
Section Two - Some Handy Short Cuts
Section Three - Literal Equations

Algebra 8 - Using Equations to Solve Problems

When you have completed this lesson you will be able to:

1) Use equations, tables, and diagrams to solve practical problems.

In order to complete this lesson successfully, you should be familiar with:

1) How to multiply polynomials.
2) How to solve first degree equations.

Section One - Guidelines for Solving Problems
Section Two - Practice in Solving Problems
Section Three - Using Tables to Organize Information
Section Four - Using Diagrams to Solve Problems

Algebra 9 - Solving Practical Problems

When you have completed this lesson, you will be able to:

1) Find formulas to describe relationships between quantities in practical problems.
2) Use formulas to organize the information in problems and to identify the variables.
3) Use formulas and organized information to develop equations.
4) Use formulas and organized information to check solutions.

In order to successfully complete this lesson, you should be familiar with:

1) Solving simple equations.
2) Using percentages.

Section One - A Problem at the Office
Section Two - A Baker's Dilemma
Section Three - Mixtures Based on Percentage
Section Four - A Money Problem
Section Five - Avoiding Trouble

Algebra 10 - Inequalities

When you have completed this lesson, you will be able to:

1) Identify inequality symbols and use them to write algebraic statements.
2) Identify the solution sets of inequalities on the number line.
3) Combine two inequalities into a compound inequality.
4) Solve inequalities.
5) Solve word problems using inequalities.

Before you begin this unit, you should:

1) Know how to locate numbers on the number line.
2) Know how to solve equations.
3) Know how to use equations to solve word problems

Section One - Inequality Symbols
Section Two - Compound Inequalities
Section Three - Solving Inequalities
Section Four - Solving Compound Inequalities
Section Five - Solving Practical Problems

Algebra 11 - Linear Equations and Graphs 1

When you complete this unit, you will be able to:

1) Identify linear equations and determine which linear equations produce vertical or horizontal lines.
2) Use graphs, equations and ordered pairs to describe the same relationship between two variables.
3) Plot points on the rectangular coordinate plane. Then use points, including the x- and y-intercepts, to graph lines on the plane.

Before you begin this unit, you should be familiar with:

1) Solving equations with one variable.
2) Graphing points on a number line.

Section One - Equations with Two Variables
Section Two - The Coordinate Plane
Section Three - Graphs of Equations
Section Four - Using the Intercepts
Section Five - Special Cases

Algebra 12 - Linear Equations and Graphs 2

When you complete this unit, you will be able to:

1) Use the term slope correctly to describe lines and rates of change.
2) Find the slope of lines on graphs.
3) Find the slope of a line when given two of its points.
4) Define the slopes of vertical and horizontal lines.

Before beginning this lesson, the student should be able to:

1) Graph ordered pairs.
2) Identify ordered pairs from a graph.

Section One - Finding Slopes from Lines
Section Two - The Slope Formula
Section Three - Negative Slope
Section Four - Finding Slope from Equations

Algebra 13 - Linear Equations and Graphs 3

When you have completed this lesson, you will be able to:

1) Use the slope-intercept form of a equation to find the slope and y-intercept of the line.
2) Write any linear equation into slope-intercept form.
3) Graph a line from its equation using the slope and the y-intercept.
4) Write an equation for a line when given the slope and y-intercept of the line.
5) Determine if two lines are parallel by comparing their slopes.

In order to successfully complete this lesson, you should be familiar with:

1) Find solutions to linear equations.
2) Using the slope formula to find the slope of a line.

Section One - Slope-Intercept Form
Section Two - Finding Slopes and Intercepts
Section Three - Using Slope-Intercept Form to Make Graphs
Section Four - Slope-Intercept Form and Equations from Graphs
Section Five - Slope-Intercept Form in Practical Problems
Section Six - Slope-Intercept vs. Standard Form

Algebra 14 - Linear Equations and Graphs 4

When you have completed this lesson, you will be able to:

1) Write an equation for a line given the slope and any point on the line.
2) Write an equation for a line given any two points on the line.
3) Write an equation for a line given the graph of the equation.

In order to successfully complete this lesson, you should be familiar with:

1) Using the slope formula.
2) Using the slope-intercept form of the equation of a line.

Section One - Writing Equations from a Point and the Slope
Section Two - Writing Equations from Two Points
Section Three - Problems with Points
Section Four - A Cost Equation
Section Five - An Equation for Price Hikes

MODUMATH ARITHMETIC CROSS REFERENCE

Basic Mathematics, Seventh Edition

MODUMATH	TEXT SECTION
1.1	1.1
1.2	1.4
1.3	1.2, 1.8
1.4	1.2
1.5	1.3, 1.8
1.6	1.5
1.7	1.5, 1.8
1.8	1.5, 1.8
1.9	1.9
1.10	1.6, 1.8
1.11	1.6, 1.8
1.12	1.8
1.13	1.7, 1.8
1.14	2.1
2.1	2.3
2.2	2.3, 2.5
2.3	3.2, 3.4, 3.5
2.4	3.2
2.5	3.3
2.6	2.6, 3.6
2.7	2.7, 3.6
2.8	6.1, 6.4
3.1	4.1
3.2	4.2, 4.3, 4.4
3.3	1.4, 4.2, 5.3
3.4	5.1, 5.5
3.5	5.2, 5.5
3.6	5.3
3.7	9.7
4.1	7.1, 7.2
4.2	7.4
4.3	7.5
5.1	11.1
5.2	11.2
5.3	11.3
5.4	11.4
5.5	11.5
5.6	11.1, 11.2, 11.3, 11.4, 11.5
5.7	---

MODUMATH ALGEBRA CROSS REFERENCE

Basic Mathematics, Seventh Edition

MODUMATH	TEXT SECTION
1	12.1, 12.5
2	12.1
3	12.1
4	---
5	12.1
6	12.2, 12.3
7	12.4
8	12.5
9	---
10	---
11	---
12	---
13	---
14	---

Instructions for using the Math Hotline

The **Math Hotline** is open 24 hours a day at 1-800-333-4227 so that students can obtain detailed hints for exercises. Exercises covered include all the odd-numbered exercises in the exercise sets, with the exception of the Skill Maintenance and Synthesis exercises.

In order to use the **Math Hotline**, students should use a <u>touch-tone</u> <u>phone</u> and dial 1-800-333-4227. Students will not be able to access the Math Hotline with a rotary phone.

They will be instructed to enter a number to select the book they're using. They should enter:

    1 for Basic Mathematics
    2 for Introductory Mathematics
    3 for Intermediate Algebra

They will then be instructed to enter the Chapter Number as a two-digit number. For example, they should enter:

    01 for Chapter 1
    02 for Chapter 2
    03 for Chapter 3, etc.

After selecting the chapter, the student will be instructed to enter the Section Number as a two-digit number. For example, they should enter:

    01 for Section 1
    02 for Section 2
    03 for Section 3, etc.

The student will then be instructed to enter the odd-numbered problem they would like assistance on as a 3-digit number. For example, they should enter:

    001 for problem number 1
    003 for problem number 3
    021 for problem number 21, etc.

After the student selects the problem number they will be given detailed hints to help them solve that problem. They can then enter the following commands to get hints to other problems:

    1 for a problem in the same section
    2 for a problem in the same chapter
    3 for a problem in the same book
    4 for a problem in a different book
    Any other command will terminate the phone call.

If you have questions about using the **Math Hotline**, please contact your local Addison-Wesley Representative.

DEVELOPMENTAL MATH PROGRAM

University of Wisconsin-Whitewater

by Phyllis Batra

The University of Wisconsin-Whitewater has a strong Academic Support Services Program. The Developmental Studies Unit falls within Academic Support Services. The Learning/Tutorial Center, Educational Opportunity Program, Minority Pre-College Program, Chicano/Latino Student Program, Upward Bound, Writing Lab, Computer Labs, McNair Post-Baccalaureate Achievement Program, and Talent Search Program are other Units working together under Academic Support Services to assist students. This article is limited to the Developmental Studies Unit.

The Developmental Studies Unit at the University of Wisconsin-Whitewater, offers academic courses in English, Reading, Basic Writing, Academic Survival/Study Skills, and Mathematics. The goal of our Unit is to provide a structured learning environment geared to increase the retention rate and grade point averages of students with one or more academic deficiencies and to enable them to maintain at least a C average for either full or part-time enrollment and eventually, GRADUATE. In addition to classes; some instructors present workshops each semester covering topics such as *Math Anxiety* or *Word Problems* which are open to the entire student body.

Developmental Math is the largest component of the Developmental Studies Unit. Our goal is to provide the at-risk mathematics student who is underprepared for basic algebra with a background necessary for success in mathematics at the college level. Prior to 1992, Developmental Math was an option offered to students. Academic advisors and counselors encouraged at-risk students to enroll in developmental courses. At that time we had 8 to 9 sections of Math each fall semester and usually 6 to 7 sections in the spring. After the Math Department (with input from developmental math staff) implemented a mandatory math placement, we increased enrollment in remedial math for the fall semester from 8 sections to 29 sections, and from 7 sections to 14 in the spring.

Proper placement in the appropriate university course is very important. Placement in a UW-W mathematics course is determined by ACT scores or the University of Wisconsin Math Placement Test (UWMPT) scores. A freshman entering the university prepared to meet the proficiency requirement in Math would enroll in 141. Those who are underprepared would enroll in 040 or 041 depending upon their test score.

Following is a copy of the letter sent to all students seeking entry at UW-W prior to their freshman orientation since mandatory placement in 1992-93. Every student is required to follow the Mathematics Placement scores stated in this letter.

THE UNIVERSITY PROFICIENCY REQUIREMENT IN MATHEMATICS

A. All students must satisfy the University Proficiency Requirement in Mathematics by fulfilling one of the following conditions:
  1. Satisfactory completion of Mathematics 760-140.
  2. Waiver from the University Proficiency Requirement in Mathematics.
  3. Granted transfer credit for any mathematics course numbered 760-140 or above (except 760-177).

B. A student who completes or receives transfer credit for 760-140 can also enroll in and receive credit for 760-141.

C. A student who completes or receives transfer credit for 760-140.

WAIVER FROM THE PROFICIENCY REQUIREMENT IN MATHEMATICS

A. A student shall be waived from the University Proficiency Requirement in Mathematics by meeting any of the following conditions:
  1. ACT each sub-score of 24 or above.
  2. SAT math sub-score of 535 or above.
  3. Score of 612 or above on AB of the UWMPT.
  4. Score of 590 or above on BC of the UWMPT.

B. A student who is waived from the University Proficiency Requirement in Mathematics cannot receive credit for 760-040, 760-041, or 760-141, but can enroll in and receive credit for 760-140.

PLACEMENT IN MATHEMATICS

Every student who has not satisfied the University Proficiency Requirement in Mathematics is required to follow the Mathematics Placement given below. This placement scheme is based on combinations of the A, B, AB, and BC scores from the UWMPT and the Arithmetic Skills Test score.

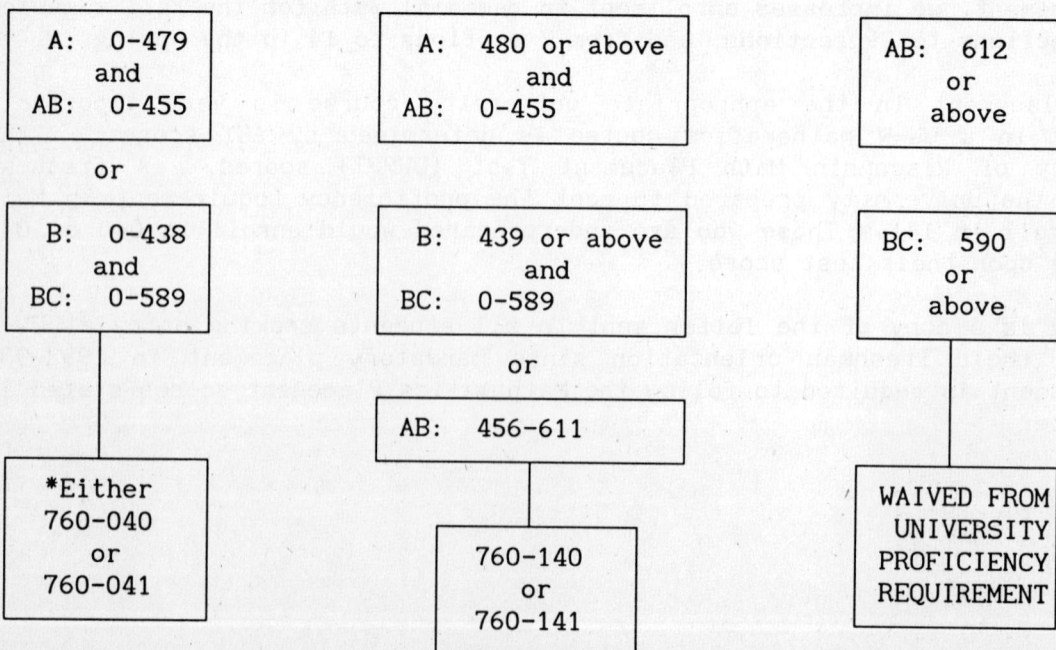

```
┌──────────────────┐        ┌──────────────────┐        ┌──────────────────┐
│  A:  0-479       │        │  A:  480 or above│        │  AB:  612        │
│      and         │        │      and         │        │       or         │
│  AB:  0-455      │        │  AB:  0-455      │        │      above       │
└──────────────────┘        └──────────────────┘        └──────────────────┘

         or

┌──────────────────┐        ┌──────────────────┐        ┌──────────────────┐
│  B:  0-438       │        │  B:  439 or above│        │  BC:  590        │
│      and         │        │      and         │        │       or         │
│  BC:  0-589      │        │  BC:  0-589      │        │      above       │
└──────────────────┘        └──────────────────┘        └──────────────────┘

                                     or

                            ┌──────────────────┐
                            │  AB:  456-611    │
                            └──────────────────┘

┌──────────────────┐        ┌──────────────────┐        ┌──────────────────┐
│  *Either         │        │  760-140         │        │  WAIVED FROM     │
│  760-040         │        │     or           │        │  UNIVERSITY      │
│     or           │        │  760-141         │        │  PROFICIENCY     │
│  760-041         │        │                  │        │  REQUIREMENT     │
└──────────────────┘        └──────────────────┘        └──────────────────┘
```

(*) Students who fall into this category must take the Arithmetic Skills Test for REQUIRED placement into 760-040 or 760-041.

Score on the Arithmetic Sills Test	Placement Required
0 - 25	760-040 then 760-041
26 - 35	760-041

After a students placement is determined they may enroll in one of the following courses.

MATHEMATICS COURSE DESCRIPTIONS AND TITLES:

Math 040 Pre-Algebra, 3 credits

A course for students who need a review of basic mathematics or who lack the computational skills required for success in algebra and other University courses. Topics include fractions, decimals, percent, descriptive statistics, English and metric units and measures of geometric figures. Emphasis is on applications. A brief introduction to algebra is included at the end of the course. This course does count toward the semester credit load and will be computed into the grade point average. It will not be included in the 120 credits required for graduation. It may be taken for a conventional grade or on a satisfactory/no credit basis.

Math 041 Beginning Algebra, 3 credits

A course for those who have a sound background in basic arithmetic, but who have not been exposed to algebra, or who need to strengthen their basic algebra skills. Topics include properties of the real numbers, linear and quadratic equations, linear inequalities, exponents, polynomials, rational expressions, the straight line, and systems of linear equations. The course counts towards the semester credit load and will be computed into the grade point average. It will not, however, be included in the credits necessary for graduation. It may be taken for a conventional grade or on a satisfactory/no credit basis.

Math 141 Intermediate Algebra, 3 credits

Introduction to college algebra. Topics and concepts extend beyond those taught in a beginning algebra course. A proficiency course for those who have not had sufficient preparation in high school to allow them to take 760-143 or 760-152.

Prereq: Satisfactory completion, with a grade of C or better, of 760-041 or demonstration of equivalent capability.

WAIVER FROM THE BASIC ALGEBRA PROFICIENCY REQUIREMENT

A student shall be waived from the basic algebra proficiency requirement (760-141) by meeting any of the following conditions:
1. ACT math score 24 or above: SAT math score of 535 or above.
2. UWMPT AB score of 612 or above.
3. UWMPT BC score of 590 or above. After the student receives placement they may enroll in other courses.

The two Developmental Math courses offered students each semester are:

Math 040 - Pre-Algebra, 3 transcript credits, Basic Arithmetic
Text used: *Basic Mathematics,* by Keedy/Bittinger, published by Addison-Wesley
Math 041 - Beginning Algebra, 3 transcript credits, Introduction to Algebra
Text used: *Beginning Algebra,* by Charles P. McKeague, Saunders College Publishing

Students who enroll in developmental math are of generic description and possess different learning styles. They arrive on campus full of enthusiasm and dreams of the future. They are wonderful human beings. They come from high schools across the state both rural and urban. Most have taken two to three math courses in high school and some are surprised and frustrated at being placed in one of the developmental classes. Others have always struggled with mathematics and are eager to improve their skills but bring with them all the old math anxieties from the past. Frequently the developmental student lacks the motivation and self-discipline to follow through keeping on task. Their self-confidence, study skills, class attendance, attention to detail and poor reading skills are some of the reasons these students fail to demonstrate strong mathematical skills needed at the college level.

A successful math program begins with good instructors who make themselves available to students. Our instructors choose to teach the developmental courses and are interested in the high-risk student. We encourage diversity in teaching to meet various student learning styles. Instructors are allowed to teach their course by whatever mode of instruction they choose and some have tried several methods over the years. The most popular method is by the lecture; followed by computer assisted instruction. Self-paced courses have worked well for some students but not all. Those students opposed to self-paced seem unusually disgruntled about this method of class structure. They often express their frustration to administrators about the lack of explicative instruction; even though the instructor outlined the course requirements clearly at the beginning of the semester and was conscientious about being available to students for math instruction and counselling in and outside of class. The students seem to dislike the burden of taking responsibility for their learning and feel the instructor is not doing enough teaching.

All instructors are required to cover the same specific material within the semester. All must distribute a detailed syllabus outlining their course, office hours, attendance, grading, and testing policies on the first day of class. All use the same standardized mid-term and final exam from those provided by the textbook companies. Instructors may use textbook exams or make up their own end of chapter or unit exams and quizzes. Most use a combination of self-made and textbook exams. They like the availability of four to six versions of each unit test to select from. Classroom space is limited and students seated near one another take different versions of the same test. This reduces the chances of cheating. Students are not allowed to keep any exams. They may go over each exam in class or by appointment with the instructor after grades are received. Students need a valid excuse to make-up or retake an exam. This is left to the discretion of the individual instructor.

Test security is of concern. Each instructor has a key to a locked file where printed standardized exams are kept. It is the responsibility of each instructor to collect all tests he or she originally hand out. Used tests are destroyed at the end of each semester.

The Learning/Tutorial Center is an important asset to our math program. The support programs occupy the entire lower level of McCutchan Hall. The facility has three classrooms, a video viewing room, two computer labs, a tutoring/study area, a Writing Lab, and the L/TC office. All appointments and record keeping are handled through the L/TC office. Math videos to accompany each chapter of the text may be checked out at the L/TC office and viewed in the Learning Center. Some videos may be checked out overnight. Math tutoring is available from 9am to 9pm, M, T, W, TR, and until 4:30pm on Fridays. No appointment is necessary for developmental math students. Tutoring is done by paid peer tutors with a strong background in mathematics and good social skills. The instructors offices are located in the same building on the second floor giving tutors and instructors the opportunity to work together to assist students. When all factors are taken into consideration; we find that the best tutorials are those requiring interaction between human beings.

The Learning/Tutorial Center assists students and instructors in Math courses by providing:

- **Walk-in Peer Tutoring**

- **Three IBM multi-media stations with laser disk drives**
  Software: Foundations for Success, Beyond Words, Math
  laser disks by Glenco/MacMillian/McGraw-Hill

- **Microcomputer assisted instruction (30 IBMs)**
  Software: Math disks by Addison-Wesley and Saunders
  College Publishing
  Open Integrated Learning System, (ILS) by MacMillian/
  McGraw-Hill
  Software by George W. Bergeman, MathCue•Tutorial,
  MathCue•Solution Finder, MathCue•Practice,
  ExaMaster+™ Computerized Test Bank, by Saunders College
  Publishing, Harcourt Brace & Company

- **Videotape Tutorials** (2 VCRs, viewing room)
  040, Basic Math, 6th edition, by Addison-Wesley, 1991
  041, Beginning Algebra, McKeague; UW-W instructor videotapes
  141, Intermediate Algebra, 6th edition, Spotlight on
  Algebra by Lial/Miller/Hornsby, Harper Collins Publisher

Our math classes are limited in size and range from 20 to 24 students in each section. This semester we will have eight 040 sections of Pre-Algebra taught by computer assisted instruction. The ILS software will be used with the optional use of the Keedy-Bittinger text as the primary source of class instruction. Instructors using ILS software exclusively at present want to incorporate one day of lecture per week next year to provide a time to teach mathematical strategies and study techniques. The other sections of 040 will be by lecture until we complete another computer lab which will be in the near future.

The 041 sections of Beginning Algebra are taught by lecture. Students have access to the instructor and walk-in assistance from peer tutors as well as videos to accompany the texts. The computer labs are networked and have appropriate ILS software available as extra practice.

We are presently in the process of setting up an additional lab of 20 computers. This will enable us to offer more math classes by computer assisted instruction. We are contemplating a way for students to take 040 outside a class, individually, on their own time. At present, it seems this is most likely to be done with the stronger remedial student or one returning to education and needing a quick refresher course.

We strive to provide an excellent instructional math program for the remedial/adult learners. We are firm with students; expecting them to take responsibility for their own learning, and at the same time offering them several means of assistance. Mastery of mathematics comes through practice and discipline. Missing class deadlines, exams, and poor attendance have penalties but we have found that they are relatively ineffective in preventing their reoccurrence with certain students.

Our most in-depth study to determine the attrition rate of students enrolled in a developmental math course was done by one of our staff members (Edna Park) in 1987. Our program had been in existence long enough to trace a significant population through their college experience. It was determined that 32% of students who enroll in Pre-Algebra are likely to graduate from the university. The normal attrition rate for college freshman is 44%. Our most recent study (1994, Park) indicated that students who completed developmental math courses were as successful in the required proficiency math course as those students who were not required to complete remedial work.

DEVELOPMENTAL STUDIES PROGRAM - REVISITED

by ANNETTE MAGYAR

MATHEMATICS COORDINATOR FOR DEVELOPMENTAL STUDIES

SOUTHWESTERN MICHIGAN COLLEGE

Here at Southwestern Michigan College in Dowagiac, Michigan, an integral part of our philosophy in the Developmental Studies department is to continually strive to serve our students better. Through monthly team meetings, inservices, conferences and professional reading we are always learning and changing to improve our eight-year old program. Even though I am involved in this process on a day-to-day basis, it was still amazing to me to actually see how much we have changed in the four years since I wrote the first article on our program for the lab manual. Now when asked to update the information I find that very little has remained the same. As I reflect on these changes I feel they have all been in our students' best interest. Certainly, we will continue to analyze our program and to improve.

At Southwestern Michigan College we have a comprehensive Developmental Studies Program housed in the Center for Academic Achievement. Our program consists of two English courses (reading and writing), English as a Second Language, four math courses and a college success strategies course. Students are either placed into this entire program or placed into only selected courses based on mandatory assessment tests administered to all incoming students. We have a dedicated and vigorous staff consisting of a director, secretary, teaching assistant, in-house counselor, curriculum specialist, three full time instructors, numerous part-time instructors and peer tutors. Since I am the only full-time math instructor in Developmental Studies, I am also the coordinator of the developmental math courses and I will expound only on the math portion of our program. The math courses in developmental studies are as follows:

        Math 099 - Mathematical Concepts (In-house text)
        Math 100 - College Arithmetic (Keedy/Bittinger text)
        Math 101A
          and   - Introductory Algebra (Keedy/Bittinger text)
        Math 101B
        Math 105 - Intermediate Algebra (Keedy/Bittinger text)

One innovation to which we have devoted much time and energy is the training of our staff in the theory and practice of cooperative learning. We had two campus wide inservices during faculty orientation which spanned several days plus biweekly follow up support meetings with our expert form Andrews University, Shirley Freed. It has been fun, frustrating and fulfilling and we are dedicated to improving and expanding our use of cooperative learning in our classes. Cooperative learning seems to be the umbrella under which our most important goals and challenges can be tackled effectively.

Our Math 099 course itself is an innovation. It is three credits, meets three hours per week, does not transfer but counts as in-house credit. This course is mandatory for those students scoring below a certain cut-off on our arithmetic placement test. It deals with those students that data has shown

125

could not pass Math 100 (College Arithmetic) no matter what extra support was offered, no matter how hard these students tried. This course concentrates on math anxiety, math study skills, learning styles and formation of basic concepts of numbers using manipulatives. There is very little computation involved. This course was developed totally in-house, based on our own in-house research and the follow up research indicates that is has been successful in meeting the needs of our department. If a student cannot pass Math 099 it is an excellent indicator that we may not be able to serve this student at this institution and career counseling follows. If a student passes Math 099 and the data shows that for the most part the student goes on to be successful in Math 100 (which is mandatory for all SMC graduates).

Math 100 (College Arithmetic) is three credits and meets three hours per week with testing outside of class time in the testing center. Although the 10 major tests and the final are in the testing center, frequent, formative quizzing goes on regularly in the classes themselves with immediate feedback and discussion to enhance learning as we go. Currently we do not allow calculators at this level. We are in the process of trying a section this summer allowing calculators about half way through after the students have sharpened their basic paper and pencil skills. We will then discuss the possibilities of adopting a new guideline for calculators and Math 100. Students must pass this course with a grade of "C" (70%) or better. This course seems to run very smoothly and successfully with homework and quizzes counting 10%, attendance 10%, tests 60% and the final exam 20%.

Math 101A and Math 101B is Introductory Algebra offered in two parts. There is an option in place for students who are able to complete the course in one semester; however, very few students are really able to do this. Each part is two credits and currently we meet only two hours per week with testing outside of class time in our testing center. The split of this course into two parts raised our successful completion rate from 50% (when it was offered in one semester only) to about 80%.

Now beginning in the Fall of 1994 we will be meeting three hours per week with an option of testing in class or in the testing room outside of class time. Each part will still be two credits. We are very excited about having this extra hour per week to implement cooperative learning lessons, error analysis, vocabulary/writing, and group problem solving to a greater degree. We believe many students will still opt to take the tests outside of class in the testing center because then they can have unlimited time. Therefore we expect the third hour per week would be available for the most part for some extremely productive activities to enhance understanding and transference of the content. Because we are predominantly a commuter school, students have little success in forming study groups outside of class so this will give them more time to work together.

Calculators are used, but not graphics calculators. Students must pass with a grade of "C" (70%) or better to be allowed to go on to the next math course. The method of computing the grade is a bit different from that in Math 100 for various reasons. We find it necessary to weigh the final a bit more and we take attendance out as a percent component at this level. Under the other plan, students could fail the final miserably and still pass the class because of good attendance. It allowed students to go on to the next math class underprepared and this was doing them a disservice. So now we will be using the following:

Homework and quizzes - 20%
                 Tests - 50%
          Final Exam - 30%

Intermediate Algebra - Math 105 is the focus of our newest project for change and improvement. After attending the American Mathematical Association of Two Year Colleges Conference in Boston this past fall, I returned with many questions and concerns involving the graphics calculator and how it fits in with our developmental math program. So we started out on a fact-finding expedition of calling, surveying, talking with people in various schools and at various levels. We have decided to first procure some comprehensive training on the graphics calculator for our entire math staff this summer with on-going monthly support meetings throughout at least the fall semester. We plan to start in the Fall of 1994 using the graphics calculator on the overhead for demonstration and to add variety and interest to our Math 105 course. Students will not be required to purchase their own as yet. We'd like to get our own feet wet first and see how it goes. When we get to the point where we are comfortable with the technology and see how our students handle it, we will decide how much, when, and where to use the graphics calculator in our curriculum. Right now my feeling is that it needs to be there in Intermediate Algebra.

Now that you have an overview of our math courses, here are some general features of our program:

Southwestern Michigan College is a community college of about 2500 students. We served approximately 970 unduplicated students in Developmental Studies in the Fall of 1993 and about 730 of these students took one of our math courses. In Winter 1993 our total unduplicated enrollment was 666. Spring 1993 was 143 and Summer 1993 was 122. As a full-time instructor in developmental studies, my classload is usually anywhere form 150-200 students total in 5 different classes plus some administrative duties as coordinator.

Our courses during fall and winter run 14 weeks and are offered during the day and evening on campus as well as at an extension sight approximately 15 miles away. During the spring and summer sessions the courses are run for 7 weeks.

Our student population is diverse in many ways. We have a wide age range, with the average being 27. We have a larger percentage of African Americans than other ethnic groups; however we do have students from all ethnic backgrounds and about 100 foreign students. These students come into our classes with a wonderful variety of attitudes, skills, learning styles and experiences.

To supplement our classroom time we have peer tutors available almost all hours of the day for one-on-one help. In general, student-tutor interaction is high and positive. All of our math tutors have either taken Math 105 and passed with an A or B or placed into a higher mathematics and are receiving an A or B in those higher math classes. In addition all of our tutors must take a one-credit training course dealing with study skills, motivation, learning styles, and teaching strategies as well as the particular details of our day-to-day operations.

The students can also sign-up for 15-minute slots with any of the math instructors during their designated office hours.

Housed in the Center for Academic Achievement are three IBM-compatible personal computers for student and faculty use. Programs from various sources are available for each level of math in our program. We also have the Addison-Wesley Keedy/Bittinger video set in the Center for use there and in the Library on a 2-day check out basis plus a set at our Extension center. These supplements are not typically used as requirements of the course; however they are used on a daily basis as optional study tools and tutoring aids and use is frequently required for students who are doing poorly on assignments or tests. Students may reserve equipment ahead of time for a one-hour block.

Auditory and visual learners do very well with the videotapes; many students find them of great help. These tapes are especially helpful if a student has missed a class, however viewing the tape does not make up the absence. The tactile/kinesthetic learners seem to enjoy the computer software as it allows them to touch and interact.

Student motivation and assuming responsibility for their own learning is our major challenge. This is partially a problem of lack of study skills, math anxiety, and lack of self-esteem or self-confidence in the school setting. But it is a much more complicated topic than it may appear on the surface. Factors such as race, socio-economic factors and many others come into play. We are continually working on strategies to address each of these aspects. Our Math 099 deals heavily with math study skills, learning styles, and math anxiety. We offer help in these areas in all our math courses but on the student's own time for the most part.

Our counselor works with learning styles, inventories, and other diagnostic tools as we refer students on an individual basis. She is also able to help students with test anxiety and time management.

Another aspect of motivation that we have declared all out war on is lack of student participation in the classroom. We are making great strides to obliterate passive student behavior by changing the climate of the classroom. We believe this is the first step to encouraging and promoting critical thinking and problem solving - another challenge that we are taking seriously. This change in our classroom climate is in part the result of the cooperative learning training we have been undergoing. By stating our expectations clearly on the first day of class and then following through, we are making progress. We are shifting from instructor as lecturer to instructor as facilitator, guide and coach. We stress attendance, that it is important that they come to class and participate. To show we care, we will call or send postcards if a student is not attending. We use intervention as a tool to inspire motivation. It's amazing what a short one-on-one intervention conference can do! I have personally had great results.

We have had successes in our Center because many students who would fall though the cracks in a more conventional setting can be helped on an individual basis. Many of those students who have been unsuccessful in their previous schooling have found new modes of learning due to varied opportunities we offer and the individual attention we give by tailoring their learning experiences to their individual needs whenever possible. Also

128

contributing to our successes is our flexibility. Students may work with any or all of the different tutors or instructors to find the best match for their learning style. Communication can be a problem in a flexible system like this so we have devised a folder sheet stapled into the inside cover of the students file on which to document any unusual conversations, assignments or arrangements between a staff member and that student, so that the next staff person to work with that student knows what has transpired. This seems to work pretty well.

We have an anecdotal file of positive student outcomes; however, we do not yet have statistics on retention, tests scores, etc. We are currently working on gathering this information.

My advice to anyone beginning a math lab or developmental studies program is:

1. To incorporate as many different teaching styles and modes of presentation as possible. No matter how clearly and thoroughly a teacher may lecture on the content, students only retain approximiately 5% of material from lecture alone. Students are not vessels into which we pour knowledge – they must talk, write, listen, read, and actively engage in their own education. I used to believe that I had to COVER all of the material in lecture and therefore didn't have time to deal with study skills, learning styles, critical thinking, group work, etc. Now I believe I can't afford not to spend time on these things because without them learning is essentially blocked. Students can get a complete lecture on video if that is what they want. My time with them can be spent more wisely. I have more to offer now especially with my cooperative learning training. Granted, it takes courage and it feels risky to deviate from the comfortable lecture, but the results have convinced me it is worth it.

2. To intervene with students as soon as there is a clue that they are struggling and intervene on a one-to-one personal level. I have had extremely gratifying fruits for this labor.

3. To hire a full-time counselor and curriculum specialist. Ideally, every program designed to serve the underprepared student needs a full-time, in-house counselor and a curriculum specialist to discover the students needs and then to find ways to meet those needs.

4. To build in adequate record-keeping and data-gathering capabilities which are essential for monitoring and evaluating the program.

THE MATH LAB

by A. David Allen

Ricks College
Rexburg, Idaho 83460

Introduction

This essay is an attempt to describe, and to some degree analyze, the characteristics of a college-level math lab. The purpose of the essay is to document the attributes of a math lab so that the first-time reader will gain some implementation ideas and encouragement for such an effort for which he or she may have responsibility. The intent of all this effort is to teach beginning mathematics successfully to the adult learner.

Overview Description

The math lab concept envisioned is a large, level-floor classroom equipped with carrels, video playback units, chalkboards, and a teacher's desk. Nearby, preferably in the same building, is a testing center with an adjacent secure, small, office-sized room for reviewing tests. Students are scheduled in the math lab for 50 minutes M, W, F, or T, Th for 75 minutes for a 15-week semester for courses in computational mathematics and algebra. The students progress on their own, through workbook-type texts with tests at the end of every chapter. Assistance is provided by tutors and the responsible faculty member. Grading is accomplished by determining the average of all chapter tests coupled with a final exam.

Characteristics of the student

Students who enroll in a math lab setting display a number of characteristics common to individuals who have done poorly in mathematics over a lifetime of education. Often they are a gregarious lot with excellent conversational skills, healthy interpersonal attributes and interest, in each other. What they often lack is solid study skills, the attention to detail that mathematics demands, and a significant attention span. At the college level they have arrived at young adulthood and beyond with significant confusion about basic computational skills and algebra. Some exhibit significant math anxiety and many are poor readers. I have on occasion had a student read aloud to me from the workbook text to assist her or him in a troublesome spot. Many have difficulty reading, and almost all have difficulty comprehending technical material. Their poor reading skills may be the reason they have reached adulthood and not learned basic mathematics.

With regard to math anxiety, I highly recommend that the reader of this note read Sheila Tobias's book, *Overcoming Math Anxiety*. She quotes Lucy Sell's work concerning women and math as the critical filter. These are very important ideas to have in mind when dealing with students in a math lab.

Pedagogy in the math lab

The pedagogical approach of the math lab is competency based. The responsibility for learning the material rests squarely on the shoulders of the student. The student's learning comes from reading the written explanations and examples in the text plus doing the calculations in the practice exercises. The presentation in the text is to follow this format:

Careful reading of Chapter 1, page 1 explanation.
Working through the authors' examples in Chapter 1, page 1.
Working margin exercises in Chapter 1, page 1.
Careful reading of Chapter 1, page 2 explanation.
Working through the authors' examples in Chapter 1, page 2.
Working margin exercises in Chapter 1, page 2.

Repeat for page 3 and so on to the end of Section 1.1. Then work 100% of the Section 1.1 exercises. Students must do all the problems in the exercise sections. Practice is one of the ingredients missing from the students' previous mathematics education.

Students should repeat the above procedure for Section 2 and so on to the end of the chapter. Next, they should complete the chapter review, and then complete the chapter test in the text. Finally, students should go to the test center and take the chapter test for grading purposes. Students can then proceed to the next chapter and repeat the above.

You might ask: Why not conduct this class using traditional faculty lecture and homework assignment methods? Are you acquainted with the problem that happens when you put these adult learners in a lecture class and attempt to teach the material at this level by the lecture method? The problem of lecturing this material like a traditional math class is that after six or seven weeks of a MWF lecture the student develops a very unhealthy dependency upon the lecturer as the fountain of all knowledge then kicks back and lets the teacher perform. As a result the focus moves from the student to the teacher and learning diminishes dramatically. Using the workbook text places the responsibility directly on the student for the entire time of enrollment, and therefore this teacher dependency never develops.

This pedagogical approach places significant responsibility upon authors, publishers, and directors of math labs to make sure the written content of the workbook text will in fact carry the instruction to the student in a complete fashion. The student *must* be able to learn from the written explanations. This is not an easy task for 100% of the students on a 100% of the topics. It is critical that the text material be pilot-tested, revised, and fine-tuned.

Student management techniques

Registration

The math lab that we are describing is not a casual, drop-in, optional operation, but one in which the student is registered for credit and enrolled in a class that meets daily M, W, F for 50 minutes, or T, Th for 75 minutes. In the college class bulletin published before every semester, the math lab is scheduled every hour from 7 A.M. to 5 P.M. and again for night classes. A listing of the class schedule for the math lab might look like Figure 1.1.

From this sample schedule, note that the student has multiple entry and exit points depending on his or her entrance skill level. One of the most challenging problems with these students is determining their initial placement in the proper class. Part of this challenge is alleviated by using our best judgment along with ACT/SAT scores and previous math classes attended. Even with these judgments, students are often enrolled in the wrong class. With the registration procedure shown in the example it is relatively easy to shift the student from one particular math program to another because all classes are going on simultaneously in the same room. Therefore in the first couple of weeks of the semester a student can proceed in the class in which she or he is registered; and if the student finds it too tough or too easy, he or she can easily make a change in registration and proceed at a different level of mathematical difficulty with no scheduling conflict.

MATHEMATICS

Code	Dept	Crs	Sec	Course Title	Start	End	Credit	Days	Time	Bldg	Rm
80813	MATH	100A	1	MATH REQ			1.0	M W F	7:00- 7:50AM	ROM	261
80814	MATH	100A	2	MATH REQ			1.0	M W F	8:00- 8:50AM	ROM	261
80815	MATH	100A	3	MATH REQ			1.0	M W F	9:00- 9:50AM	ROM	261
80816	MATH	100A	4	MATH REQ			1.0	M W F	10:00-10:50AM	ROM	261
80817	MATH	100A	5	MATH REQ			1.0	M W F	12:00-12:50PM	ROM	261
80818	MATH	100A	6	MATH REQ			1.0	M W F	1:00- 1:50PM	ROM	261
80819	MATH	100A	7	MATH REQ			1.0	M W F	2:00- 2:50PM	ROM	261
80820	MATH	100A	8	MATH REQ			1.0	M W F	3:00- 3:50PM	ROM	261
80821	MATH	100A	9	MATH REQ			1.0	M W F	4:00- 4:50PM	ROM	261
80822	MATH	100A	10	MATH REQ			1.0	T T	8:00- 9:15AM	ROM	261
80823	MATH	100A	11	MATH REQ			1.0	T T	9:30-10:45AM	ROM	261
80824	MATH	100A	12	MATH REQ			1.0	T T	11:00-12:15PM	ROM	261
80825	MATH	100A	13	MATH REQ			1.0	T T	12:30- 1:45PM	ROM	261
80826	MATH	100A	14	MATH REQ			1.0	T T	3:00- 4:15PM	ROM	261
80827	MATH	100A	190	MATH REQ			1.0	W	7:00- 9:40PM	ROM	261

(Students registered for MATH 100B must attend the first class to be guaranteed a seat.)

Code	Dept	Crs	Sec	Course Title	Start	End	Credit	Days	Time	Bldg	Rm
80829	MATH	100B	1	BEG ALGEBRA I			1.0	M W F	7:00- 7:50AM	ROM	261
80830	MATH	100B	2	BEG ALGEBRA I			1.0	M W F	8:00- 8:50AM	ROM	261
80831	MATH	100B	3	BEG ALGEBRA I			1.0	M W F	9:00- 9:50AM	ROM	261
80832	MATH	100B	4	BEG ALGEBRA I			1.0	M W F	10:00-10:50AM	ROM	261
80833	MATH	100B	5	BEG ALGEBRA I			1.0	M W F	12:00-12:50PM	ROM	261
80834	MATH	100B	6	BEG ALGEBRA I			1.0	M W F	1:00- 1:50PM	ROM	261
80835	MATH	100B	7	BEG ALGEBRA I			1.0	M W F	2:00- 2:50PM	ROM	261
80836	MATH	100B	8	BEG ALGEBRA I			1.0	M W F	3:00- 3:50PM	ROM	261
80837	MATH	100B	9	BEG ALGEBRA I			1.0	M W F	4:00- 4:50PM	ROM	261
80838	MATH	100B	10	BEG ALGEBRA I			1.0	T T	8:00- 9:15AM	ROM	261
80839	MATH	100B	11	BEG ALGEBRA I			1.0	T T	9:30-10:45AM	ROM	261
80840	MATH	100B	12	BEG ALGEBRA I			1.0	T T	11:00-12:15PM	ROM	261
80841	MATH	100B	13	BEG ALGEBRA I			1.0	T T	12:30- 1:45PM	ROM	261
80842	MATH	100B	14	BEG ALGEBRA I			1.0	T T	3:00- 4:15PM	ROM	261
80843	MATH	100B	190	BEG ALGEBRA I			1.0	W	7:00- 9:40PM	ROM	261

(Math 100C requires consent of Instructor. May begin as soon as MATH 100B is completed.)

Code	Dept	Crs	Sec	Course Title	Start	End	Credit	Days	Time	Bldg	Rm
80860	MATH	100C	190	BEG ALGEBRA II			1.0	W	7:00- 9:40PM	ROM	261
80845	MATH	100C	301	BEG ALGEBRA II	10/20	12/16	1.0	M W F	7:00- 7:50AM	ROM	261
80846	MATH	100C	302	BEG ALGEBRA II	10/20	12/16	1.0	M W F	8:00- 8:50AM	ROM	261
80847	MATH	100C	303	BEG ALGEBRA II	10/20	12/16	1.0	M W F	9:00- 9:50AM	ROM	261
80848	MATH	100C	304	BEG ALGEBRA II	10/20	12/16	1.0	M W F	10:00-10:50AM	ROM	261
80849	MATH	100C	305	BEG ALGEBRA II	10/20	12/16	1.0	M W F	12:00-12:50PM	ROM	261
80850	MATH	100C	306	BEG ALGEBRA II	10/20	12/16	1.0	M W F	1:00- 1:50PM	ROM	261
80851	MATH	100C	307	BEG ALGEBRA II	10/20	12/16	1.0	M W F	2:00- 2:50PM	ROM	261
80852	MATH	100C	308	BEG ALGEBRA II	10/20	12/16	1.0	M W F	3:00- 3:50PM	ROM	261
80853	MATH	100C	309	BEG ALGEBRA II	10/20	12/16	1.0	M W F	4:00- 4:50PM	ROM	261
80854	MATH	100C	310	BEG ALGEBRA II	10/20	12/16	1.0	T T	8:00- 9:15AM	ROM	261
80855	MATH	100C	311	BEG ALGEBRA II	10/20	12/16	1.0	T T	9:30-10:45AM	ROM	261
80856	MATH	100C	312	BEG ALGEBRA II	10/20	12/16	1.0	T T	11:00-12:15PM	ROM	261
80857	MATH	100C	313	BEG ALGEBRA II	10/20	12/16	1.0	T T	12:30- 1:45PM	ROM	261
80858	MATH	100C	314	BEG ALGEBRA II	10/20	12/16	1.0	T T	3:00- 4:15PM	ROM	261
*80861	MATH	101	1	INTER ALGEBRA			3.0	M W F	7:00- 7:50AM	ROM	261
*80862	MATH	101	2	INTER ALGEBRA			3.0	M W F	8:00- 8:50AM	ROM	261
*80863	MATH	101	3	INTER ALGEBRA			3.0	M W F	9:00- 9:50AM	ROM	261
*80864	MATH	101	4	INTER ALGEBRA			3.0	M W F	10:00-10:50AM	ROM	261
*80865	MATH	101	5	INTER ALGEBRA			3.0	M W F	12:00-12:50PM	ROM	261
*80866	MATH	101	6	INTER ALGEBRA			3.0	M W F	1:00- 1:50PM	ROM	261
*80867	MATH	101	7	INTER ALGEBRA			3.0	M W F	2:00- 2:50PM	ROM	261
*80868	MATH	101	8	INTER ALGEBRA			3.0	M W F	3:00- 3:50PM	ROM	261
*80869	MATH	101	9	INTER ALGEBRA			3.0	M W F	4:00- 4:50PM	ROM	261
*80870	MATH	101	10	INTER ALGEBRA			3.0	T T	8:00- 9:15AM	ROM	261
*80871	MATH	101	11	INTER ALGEBRA			3.0	T T	9:30-10:45AM	ROM	261
*80872	MATH	101	12	INTER ALGEBRA			3.0	T T	11:00-12:15PM	ROM	261
*80873	MATH	101	13	INTER ALGEBRA			3.0	T T	12:30- 1:45PM	ROM	261
*80874	MATH	101	14	INTER ALGEBRA			3.0	T T	3:00- 4:15PM	ROM	261
*80875	MATH	101	15	INTER ALGEBRA			3.0	M W F	4:00- 4:50PM	ROM	260
*80876	MATH	101	190	INTER ALGEBRA			3.0	T	7:00- 9:40PM	ROM	261

Figure 1.1
Math Lab Schedule

Student Progress

When a class organization permits extensive latitude for the student in the area of self direction, it is important that the student be provided a schedule of minimum progress. I have claimed that this pedagogy I have been describing is competency based. It would be more accurate to refer to the program as a "modified" competency-based system. In a truly competency-based program, like the one Keller suggested in the early 1970's for college physics classes, we would establish a high standard--90%, for instance--for the tests, and a student could not progress to the next chapter until the 90% level was reached. Grading under a truly competency based system is either "A" or "not yet completed." That system has some distinct disadvantages at the end of the term, and one of the major corrections needed is to provide the student with a minimum progress schedule and modified grading scale based on test average.

Each student is provided a personal copy of the schedule of the class for which he or she is enrolled. The instructor also has a copy of each student's schedule. During the course of the classroom period the instructor can check with each student concerning his or her progress and then make a notation on the instructor's copy of the student's schedule. The instructor should compile a three-ring binder with a schedule page for each student.

Figure 1.2 is a sample minimum progress schedule.

Student Information Sheet

During the first few minutes of class on the first day the following information sheet can be provided to the student. This sheet will answer questions about placement, attendance, homework, tests, grades, available help, and lab rules. It is very important that the student have this material in written form so that future reference can be made to the outline.

The following pages are the information sheet:

ATTENDANCE AND SCHEDULE

Major _____   Name _____
I.D.# _____   Seat # _____
Phone # _____

Aug 30 - 1.1, 1.2 ____	Oct 23 - 6.8 ____
31 - 1.3, 1.4 ____	24 - 6.8 ____
Sep 1 - 1.5, 1.6 ____	25 - 6.9 ____
	26 - 6.10 ____
4 - LABOR DAY ____	27 - TAKE TEST 6 ____
5 - 1.7, 1.8 ____	
6 - TAKE TEST 1 ____	30 - 7.1 ____
7 - 2.1, 2.2 ____	31 - 7.2 ____
8 - 2.3, 2.4 ____	Nov 1 - 7.3 ____
	LAST DAY TO RETAKE 6
11 - 2.5, 2.6 ____	2 - 7.4 ____
LAST DAY TO RETAKE 1	3 - 7.5 ____
12 - 2.7 ____	
13 - 2.8 ____	6 - 7.6 ____
14 - TAKE TEST 2 ____	7 - 7.7 ____
15 - 3.1, 3.2 ____	8 - 7.8 ____
	9 - 7.9 ____
18 - 3.3 ____	10 - 7.10 ____
19 - 3.4 ____	
LAST DAY TO RETAKE 2	13 - 7.11 ____
20 - 3.5 ____	14 - 7.11 ____
21 - TAKE TEST 3 ____	15 - TAKE TEST 7 ____
22 - 4.1, 4.2 ____	16 - 8.1 ____
	17 - 8.1 ____
25 - 4.3, 4.4 ____	
26 - 4.5, 4.6 ____	20 - 8.2 ____
LAST DAY TO RETAKE 3	LAST DAY TO RETAKE 7
27 - TAKE TEST 4 ____	21 - 8.3 ____
28 - 5.1, 5.2 ____	22 - 8.3 ____
29 - 5.3, 5.4 ____	23 - THANKSGIVING
	24 - THANKSGIVING
Oct 2 - 5.5 ____	
LAST DAY TO RETAKE 4	27 - 8.4 ____
3 - 5.6 ____	28 - 8.5 ____
4 - 5.7 ____	29 - 8.6 ____
5 - 5.8 ____	30 - 8.7 ____
6 - 5.9 ____	Dec 1 - 8.8 ____
9 - 5.10 ____	4 - 8.8 ____
10 - 5.11 ____	5 - 8.9 ____
11 - TAKE TEST 5 ____	6 - 8.9 ____
12 - 6.1 ____	7 - TAKE TEST 8 ____
13 - 6.2, 6.3 ____	8 - Review ____
16 - 6.3 ____	11 - Review ____
LAST DAY TO RETAKE 5	LAST DAY TO RETAKE 8
17 - 6.4 ____	13-15 Final
18 - 6.5 ____	
19 - 6.6 ____	
20 - 6.7 ____	

Test 1 ___,___,___,___,___.
Test 2 ___,___,___,___,___.
Test 3 ___,___,___,___,___.
Test 4 ___,___,___,___,___.
Test 5 ___,___,___,___,___.
Midterm grade _____

Test 6 ___,___,___,___,___.
Test 7 ___,___,___,___,___.
Test 8 ___,___,___,___,___.
Average ____
Final Exam Score ____
Final Grade ____

Figure 1.2
Minimum Progress Schedule

MATH LAB STUDENT INFORMATION SHEET

Welcome to the Math Lab.

There are some things that you need to understand about Math 101. Please take time to read this information carefully.

First, a large number of students are enrolled in the Math Lab. Attendance is required, and we feel that those students attending should be able to work in an atmosphere conducive to study and thought. We expect that all students will work quietly. A student may be asked to leave by the instructor or a tutor (with an absence recorded) if he/she persists in disturbing classmates.

Second, let's be sure that you are enrolled in the correct course. A brief description of Math 101 is:

MATH 101--INTERMEDIATE ALGEBRA--(semester course, 3 credits)

The prerequisite for Math 101 is Math 100B-C or Algebra I. It is expected that the student has a working knowledge of basic algebraic concepts. Having been enrolled in Algebra I in high school does not necessarily indicate the necessary background for success in Math 101. The textbook material covered in this course is Chapters 1 - 8 in **Intermediate Algebra** by Keedy and Bittinger.

A grade of C or above in Math 101 fulfills the math requirement. A grade of C- **will not** fulfill the requirement.

Once you have decided this is the course that best meets your needs, you should buy the appropriate textbook and begin. The courses in the Math Lab are self paced. This means that you work on your own, at your own pace. You will receive a schedule that suggests the pace at which you should proceed. You need to stay on this schedule in order to finish the course within the alloted time. You may go faster than the schedule suggests, but you should not go at a slower pace.

Math 101 students will begin with the first page of Chapter 1. Work through the chapter, doing all of the margin exercises and the odd numbered exercises for each section (some instructors will require all the exercises). When you feel you understand the material, do the review test at the end of the chapter. Having done this successfully, go to the Testing Center and ask for the appropriate test. You will work through each chapter in this way until the course work is completed.

If at some time you feel that you are not enrolled in the correct course, see your instructor. It is possible to change your registration. If you have any doubt as you begin, it is suggested that you do not write in your textbook the first few days.

Your grade depends on your attendance, homework, and test scores, so let's learn more about them.

Figure 1.3
Student Information Sheet P. 1

ATTENDANCE

To be given credit for attendance, you are expected to arrive in class on time and to remain in your assigned seat, working the full class period. If you come in late or leave early, you may be marked absent for the class.

Attendance is required every day. We feel that you need to be in class to make proper progress.

The policy is:

> 5 absences -- No penalty
> 6-9 absences -- Grade dropped one letter grade
> over 10 absences -- An F will be given (student should drop if it is possible)

All absences for personal reasons are unexcused. It is expected that five absences should be enough for illness, emergencies, trips home, etc. Be sure that you are absent only when it is absolutely necessary.

Absences due to class field trips will be excused if you show your instructor the appropriate form, signed by the person in charge of the activity.

You will be expected to attend class until you have taken the final exam and have shown the results to your instructor.

HOMEWORK

You will be required to do the odd numbered problems in the exercises as homework (some instructors may require all the problems). The homework problems will be those in the sections scheduled for the day and will be done at the beginning of the hour the next day. You must show your work to earn credit. Not doing homework will result in a lower grade--possible a failing grade.

TESTS

All tests are taken in the Testing Center. You must have your student I.D. card and be properly registered before you can take a test.

After taking a test in the Testing Center, you will be given a printout showing how you did on that test. Show this form to your instructor and then keep it until you receive your grade. It is your responsibility to verify your test scores if necessary.

For Math 101 students there will be eight chapter tests and a final. Each chapter test will have 25 problems; 5 of these problems will be taken from the preceding chapters. Algebra tests are in the Math Testing Center for a limited time, and the closing dates are listed on the class schedules. You are expected to take the chapter tests **on** or **before** the date listed on the class schedule. If you do not take the test as scheduled, you will receive a 0 on the test.

Retakes may be taken until the closing date, but **you may not take a chapter test after the "last day."**

Calculators are allowed on Math 101 tests.

Figure 1.4
Student Information Sheet P. 2

TEST RETAKES

If an algebra student is not satisfied with a chapter test score, it is possible to retake a test. A fee will be charged by the Testing Center for these retakes.

Before you retake a test, see the sample tests posted in the Lab. For example, if you missed numbers 3, 5, and 7 on a given test, numbers 3, 5, and 7 on the sample test will show you the types of problems you missed.

The Math 101 final may be take **one time only**.

NOTES ABOUT THE TESTS

The tests have been printed on a computer. Thus, $\frac{1}{2}$ is written as 1/2 and $a + \frac{b}{c}$ as a + b/c.

It has been necessary to make some changes in black ink on the tests, so it is important that students do not make additional marks on the tests.

GRADES

For Math 101, grades are determined by the **average of the scores of all the tests you take**. For example, if you take the Chapter 1 test three times, the average of those three test scores will be used as your Chapter 1 score. Any Chapter test not taken is averaged in as a zero.

The test average on all eight chapter tests determines your grade before the final:

96-100	90-95	80-89	77-79	73-76	70-72	66-69	60-65	50-59
B+	B	B-	C+	C	C-	D+	D	D-

The score you receive on the final will determine whether or not you receive a higher grade than indicated on the chart.

Score on final:

        90%-100%--raise one full grade (for example, B to A)
        80%-89%--raise two steps (for example, B to A-)
        70%-79%--raise one step (for example, B to B+)
        Below 70%--grade remains the same

Example: A student has an average of 83 and receives a score of 75 on the final. The 83 gives a grade of B-; the 75 on the final raises the grade one step to a B.

An A will be the highest grade given.

Figure 1.5
Student Information Sheet P. 3
138

INCOMPLETE GRADES - Incomplete grades are rarely given out. The incomplete grade policy is as follows: An I is given only when extenuating circumstances (serious illness, death in the immediate family, etc.) occur after the twelfth of the semester. If the extenuating circumstances arise before the twelfth week of a semester, the student should petition through the Registrar's Office to drop the class.

You must have a grade of **at least** B before you take the next math course in your sequence. Any grade less than this indicated you have not mastered the material well enough to complete the next course successfully.

WHERE CAN YOU GET HELP?

The tutors and your instructor are available to help you during class.

Your instructor can be seen outside of class.

Your instructor is: _____

Office and phone: _____

Office hours are: _____

Videotapes are available for your viewing. Most sections in the textbooks have a corresponding videotape lecture on the material covered in that section. They are located in the Library, 3rd floor video viewing area or in Room 261.

Computer Programs are available as an added method of learning algebra. Math Lab software for Math 101 can be obtained from your instructor for use in Room 261. See the topic listings posted on the bulletin board.

Sample tests are posted on the bulletin board in the Lab. The numbers and types of problems on the sample tests correspond with those on the actual tests. These sample tests are also available on reserve in the library, where they may be copied.

Aids available in the reserve section of the library are:

Sample tests in booklet form.
Student's Solutions Manual for Intermediate Algebra (this booklet shows the solutions for odd numbered problems).
Even answers for **Intermediate Algebra**.
Extra practice sets in booklet form for the text.

SOME MATH LAB REMINDERS

1. Please do not bring food or drink into the lab.

2. Please work quietly, being considerate of those working around you.

WE WISH YOU SUCCESS IN THE MATH LAB. IF YOU HAVE QUESTIONS, ASK. REMEMBER THAT WE ARE HERE TO HELP YOU!

Figure 1.6
Student Information Sheet P. 4

The Test Center and Testing

A math lab program like the one we have described requires the availability of a test center that is open for extended evening and Saturday hours. The test center must be a secure operation--one in which the employees are trustworthy. The test center that functions well has automated equipment to score, record, and summarize test results from multiple-answer tests. The faculty members responsible for the test center will need to prepare a testbank to parallel the workbook and text the students are using. It will be necessary to prepare chapter tests and a final. The tests should be prepared in five forms to minimize student collaboration and provide retake possibilities. The student may take any form as the initial test, and the other four forms then act as retake possibilities. The score for any chapter is determined to be the average of all forms taken for that chapter.

The test center must have a report program to keep the responsible faculty members advised of student progress and a reporting system to provide feedback to the student. The next pages include the following:

    a multiple-choice test
    an answer sheet completed by a student (answers 17 and 18 are wrong)
    an individual report to the student for this answer sheet
    a test center weekly report

A private, office-sized room must be available adjacent to the test center so that during specific hours of the day a tutor can review specific test questions with students. The purpose of this activity is to provide feedback to the student on her or his errors so that when the student retakes the exam she or he does not make the same types of errors again.

MATH 101   TEST 5 C   PAGE 1

DO NOT WRITE ON THIS TEST

1. Collect like terms: $4ab + 7a^2b^2 - 4a + 2ab - 3a^2b^2$.
   (a) $10a^2b^2 + 6ab - 4a$ (b) $4a^2b^2 - 4a + 6ab$
   (c) $10a^3b^3 - 4a$ (d) $6a^2b^3$
   (e) None of these

2. Add: $x^3y + 3x^2y - 3xy + 3y$ and $13x^3y + 4x^2y - 8y$
   (a) $13x^4y$ (b) $14x^3y + 7x^2y - 8xy^2$
   (c) $14x^3y + 7x^2y - 11y$ (d) $14x^3y + 7x^2y - 3xy - 5y$
   (e) None of these

3. Subtract: $(2x^3 - 3x^2 + 5) - (2x^2 - x - 3)$
   (a) $2x^3 - 5x^2 + x + 8$ (b) $2x^3 - 5x^2 - x + 2$
   (c) $2x^3 + 2x^2 + x + 8$ (d) $-3x^2 - x + 8$
   (e) None of these

4. Multiply and simplify: $8y^2$ and $7y^3$
   (a) $54y^5$ (b) $54y^6$
   (c) $56y^6$ (d) $56y^5$
   (e) None of these

5. Multiply: $12m$ and $3m - 8$
   (a) $36m^2 - 72m$ (b) $15m - 8$
   (c) $36m^2 - 96m$ (d) $36m^2 - 8$
   (e) None of these

6. Multiple: $7x^2 + 3$ and $-x - 4$
   (a) $7x^3 - 28x^2 - 3x -12$
   (c) $7x^3 - 3x^2 - 12$ (b) $7x^3 - 28x^2 - 3x + 12$
                      (d) $7x^3 - 31x2 + 12$
   (e) None of these

7. Factor completely: $3x^2y^2 - 9xy^2 - 3xy$
   (a) $-3xy(xy - 3y + 1)$ (b) $3xy(xy - 3y - 1)$
   (c) $-3xy(-xy - 3y - 1)$ (d) $3xy(xy - 3y + 1)$
   (e) None of these

Figure 1.7
Test, Multiple Choice (only 1 page)
140

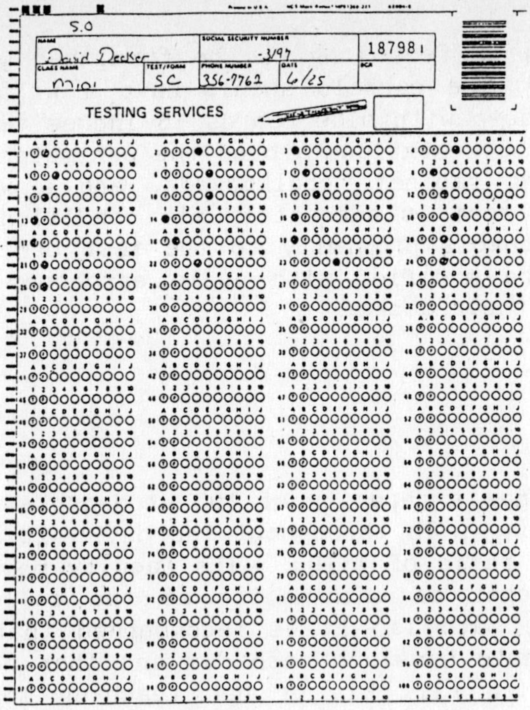

Figure 1.8
Student Answer Sheet

INDIVIDUAL REPORT

DECKER DAVID EARL 3197

INTER ALGEBRA MATH 101-601 11:54:07 25 JUN 1990

Test: 5 Form: C Percent: 92 Score: 23

Entry Time: 11:06 A.M. Elapsed Time: 00:47 Answer Sheet: 187981

ITEMS MISSED: 017 018

Report to Student

Figure 1.9
Report to Student

* * * * * * * * * * * * * * * * SECTION 601 * * * * * * * * * * * * * * * *

TESTING SERVICES--S.C.O.U.T. - INTER ALGEBRA - MATH 101

COLUMN / ID NUMBER	-01- TST1	-02-	-03- TST2	-04-	-05- TST3	-06-	-07- TST4	-08-	-09- TST5	-10-	-11- TST6	-12-	-13- TST7	-14-	-15- TST8	-16-	-17- FINL	-18- PRCT
2025	90		96															23
0816	92		100															24
3773	88		86															22
3998	87		96															23
0412	92		88		0													23
5667	100		94															24
6172	72																	9
7449	72																	9
5910	84																	11
3197	96		92		80													34
9746	92		100															24
5304	89		88															22
0106	92		96		96		96											48
0144	84		0															11
2605																		
9233	84		88															22
6482	80		92		72													31
1755	92																	12
7873	84		64		48													25
0892	84		42															16
6634	64		80															18
2317	96		92		0													24
0220	90		84		88													33
3666	88																	11
7494	92		100		0													24
9776	76																	10
8963	92		92															23
0712	88		92															23
5288	84		92															22
6004	92		88															23
9987	84		96															23
8223	88		84															22
0400	92		80															22
5513	88		100															24
4858	80		82															20
MAXIMUM SCORES.......	25		25		25		25											200
MEANS................	86.7		88.3		76.8		96.0											21.6
STANDARD DEVIATIONS..	7.5		12.2		18.5		0.0											7.9
HIGHEST SCORES.......	100		100		96		96											48
LOWEST SCORES........	64		42		48		96											9
COUNT OF TESTS GIVEN.	34		27		5		1											34

Figure 1.10
Report to Instructor

Physical facilities

A level-floored classroom, 40 ft wide by 50 feet long can be furnished with 100 carrels with pedestal-type seating. The carrels should have a writing surface of 30 inches wide and 20 inches deep with side boards 15 inches tall. This arrangement will allow 10 video playback stations around the perimeter of the room and a teacher's desk, a filing cabinet, chalkboards, and a wall cabinet. It will be essential that the room is adequately air conditioned. This facility should be located in reasonable proximity to the test center. The 10 video playback machines should be modified with audio earphones to eliminate disturbing students at work at their desks.

Staffing

Faculty

The math lab facility we are describing has a capacity of 100 students every 50 minutes on M, W, F and every 75 min on T, Th throughout the day. It has worked well to have a faculty to student ratio of 1:100. The faculty member should be a full-time professional teacher of mathematics at the college level. Because of the intensity of the math lab it is recommended that a teacher conduct at most two 50 or 75-minute sessions each day and then have other, traditional, mathematics courses to fill his or her teaching load. Other teachers in the department can also take their turn in the math lab. It has been determined that this 1:100 ratio is a maximum number that one faculty member manage in order to visit each student's carrel in the 50 or 75-minute period to check progress, encourage students who are slacking, praise students who are on schedule, and make a notation in the three-ring binder that contains a schedule page for every student. Even though there is a brief meeting in many cases between student and teacher, it is sufficient to convey interest in the student, encourage his or her progress, answer a question, and thus motivate the student to action.

Student Tutors

Student tutors in the math lab are also essential. It has been determined that a student to tutor ration of 20:1 be established. The tutors are students who have been successful at mathematics and have earned an A grade in college algebra or some other more advanced college-level mathematics course. In a class of 100 students there would then be six individuals working in the classroom--five student tutors and the faculty member.

Student Tutor Training

Student tutors should receive an orientation and training.

Mathematics Tutoring

Introduction

Tutoring is a very intense, one-on-one experience that requires effort and skill to be of value in the life of the student. Many personal characteristics as well as skill in the academic discipline are required to assist a student who needs help. In an effort to be complete in our orientation of tutors the following ideas have been noted with explanation.

Work Habits

   Punctuality--Tutors need to arrive early to class to get material out and present an attitude of interest as students arrive.

   Patience

   Enthusiasm--A tutor should convey a sense of importance and need through his conversation and explanations to the student. A sustained effort in this area will be perceived as enthusiasm for mathematics.

   Honesty

   Slow to anger

   Listening--A tutor needs to listen very carefully to a student's explanation of a mathematical process so that any errors in thinking can be corrected.

   Hard work

   Willingness

   Maintaining academic standards--The tutor must remember that he or she is to follow instructions as laid out by the faculty member and should not modify any policy. Pressure from students must be withstood.

Explaining ability

In mathematics there are many tricky areas, such as common denominators for algebraic fractions, radicals, percent, and others, that require practice in proper presentation to the student. The tutor will need to make very clear explanations in troublesome areas. In addition to verbal skills in explanations tutors must have chalkboard writing skills that are much above average.

Teaching ability

Tutors are teachers and need to develop the skills of good teaching, such as explanation, questioning, responding to questions, sensing when learning is taking place, analysis of problems, and providing alternative example problems.

Attitude

With her or his responses, the tutor can set the stage for student feelings about a course. If a tutor becomes cavalier about learning, he or she will convey to the student feelings that will interfere with learning.

Dealing with student frustration

Mathematics will heighten student frustration as quickly as any discipline. Tutors need to be able to sense these kinds of feelings and get the student working productively again. This is usually best done by backing up and looking with the student at the content, determining the place where understanding is solid, and proceeding again through the content in a methodical step-by-step procedure.

Mannerisms that encourage students to ask for assistance

Tutors moving quietly around the room--This aids students in getting easier access to the tutor.

Asking students how they are doing--On occasion a nonintrusive question like, "How are you doing?" will lead to a productive exchange between student and tutor.

Responding to unasked questions--Watching the class carefully and sensing when a student needs help, for example, by the way he fidgets in his chair.

Mannerisms that discourage students from asking for assistance

Sitting on a desk or chair.

Looking at the floor or out the window. Generally not paying attention to the class and as a result missing a raised hand.

Carrying an "air of stand-off-ishness" or "I am more important than you" that students can sense and that will inhibit their desire to ask questions.

Ethnic Differences

It has been suggested that the following tutoring techniques may be offensive to some students:

looking them in the eye

female tutors assisting male students

asking male students "do you understand?"

touching the student

Dating

Dating and personal interest between tutors and students may cause some difficult problems.

Friendship among tutors

Tutors need to remember that if they are tutoring together during the same class, a gab fest is out of order during slack moments.

Personal grooming for men tutors--areas of concern:

Haircuts

Body odor

Fingernails and hands

Unshined shoes

Bad breath

Smile

Personal grooming for women tutors--areas of concern:

Unkempt hair

Body odor

Blouse and dresses wide open at the neck

Thin, sheer, and tight-fitting clothing

Fingernails and hands

Bad breath

Smile

Textbook and Supplementary Material

To accomplish this type of mathematical presentation, the publisher and the author must supply a program, not simply a book. That program must include the following five items:

1. Textbook with: explanations
                     margin exercises
                     section exercises
                     chapter review with problems keyed to explanations
                     sample tests with problems keyed to explanations
                     answer key for all problems keyed to explanations
                     answer key with no errors. (This is tough to do for first edition texts. If there are many errors the students really get discouraged.)

2. Testbank booklet with five forms of every chapter test. The testbank must be in multiple choice format so that the instructor can either use them as is or use the tests without the multiple answers as a completion test.

3. Reference booklet of all section exercises completed with every step of of the calucations shown.

4. Videotape of an instructor lecture for every section in the text. Lectures should not exceed 20 minutes per section, and two or three lectures can be put on one tape.

5. Instructors manual with this material.

Mathematics by its nature is tough for students. They need help and explanation plus good graphics in the text to learn this material. We cannot condense material per page to the point that is too much for the student. One of the problems that all authors and publishers have with the worktext concept is the presentation of an entire page of math calculations that hits the eye of the student when the student turns the page. Recall that in the classroom the chalkboard is empty when we start. The student therefore gets the chance to see the "flow" and "dynamics" of a problem being solved and to see the solution built line by line. In this way, the problems don't appear quite so imposing as an instantaneous page full of calculations. To accomplish the intended outcome the author's writing style should be direct, carefully defined, and descriptive, with no content gaps, and examples should be rich with complete, not sketchy, explanations. The student's point of view should be that he or she can understand this material and that the text is not overwhelming just when he or she looks at the pages.

As I look at the examples and exercise sets, the question in my mind is whether the examples are carried out in sufficiently small steps and whether they build on each other logically without content gaps or deletions so as to carry the instruction along in the mind of the student with continuity. Since the student is doing the work in an independent fashion without the benefit of a live lecture-explanation, this is critical. The acid test for this issue is "Has the text been pilot-tested through enough editions to serve the student's needs?" As I have used worktexts it becomes so obvious where the gaps are located. The gaps are where significant numbers of students stumble, hands go up, and tutors are busy bridging the gap with one-on-one explanations. To determine accurately where these gaps occur in the text explanation, keep a classroom log of where the troublesome spots occur. Then make suggestions for improvement of the next edition with the mathematics editor of the publishing company.

Issues for College Administrators to consider:

Does the college commit itself to assisting the adult learner in the areas of computation mathematics and algebra, or is this considered to be remedial material that should have been learned in grade school and junior high school and therefore not to be bothered with?

Is the college willing to commit $20 per hour for a tutoring budget and $20 per hour for a faculty salary budget for a math lab of a capacity of 100 every hour?

Is the college willing to commit 2000 square feet of space and one-time furnishings budget of $150 per student to provide a facility for a 100-seat math lab, plus the usual needs of heat, air conditioning, and custodial services?

Is the college registrar willing to process double the normal number of registration changes to accommodate the students in this type of mathematics instruction?

Does the college have a campus-wide testing center that operates on extended hours (7 A.M. to 9 P.M.) and Saturdays?

Conclusion

Mathematics, music, and foreign language are the real disciplines of life. As teachers of mathematics, it is imperative that we provide the best instructional design to programs of development mathematics for adult learners as would schools of language or music.

One school of thought suggests that arithmetic and algebra are developmental and are to be learned in a school year of 185 days, in 50-minute periods each, when the student is 13, 14, or 15 years of age. If there is even a particle of truth to that statement our efforts at the adult level need to be of such a quality to really motivate, instruct, and evaluate students who enroll in our programs.

The mathematics lab described in the preceding pages is a description of the sustained effort each semester since the Fall of 1974 to the Summer of 1994. It works for most students. It attends to many styles of learning that students exhibit. It is so refreshing to step into the math lab two minutes after the starting time and see so many students hard at work without the normal call to order of the professor in the traditional lecture method. The students do feel the responsibility is directly on their shoulders and proceed each day with no direct instruction from the teacher. Most students gain the confidence they need to proceed with the next math class in their program.

For addition information: A. David Allen
                              Math Dept.
                              Ricks College
                              Rexburg, Idaho 83460

                              Office: 1-208-356-1926
                              Home: 1-208-356-9063

The Mathematics Learning Center at New Mexico State University

Kitty Berver

Background
New Mexico State University (NMSU) is a comprehensive land-grant institution that originally had open admission standards and has only recently instituted minimal admission standards. Approximately 35% of the 16,000 students on the main campus are from a minority group, primarily Hispanic (30%) and Native American (3%). The math lab, the Mathematics Learning Center (MLC), was established in 1979 to teach non-credit developmental courses in arithmetic and basic algebra. Two-thirds of the students enrolled in the non-credit sequence had graduated from high school less than two years previously. Reteaching the material in the same fashion as the high school course did not seem to be an adequate solution. The MLC design sought to ensure that students who completed these courses did not have the same gaps in their knowledge that had initially prevented them from comfortably using arithmetic and algebra skills.

The program was modeled after the learning strategy entitled "Personalized System of Instruction" (PSI) authored by Fred S. Keller. It was gradually extended to include intermediate algebra, college algebra, and trigonometry. Initially the department offered both traditional lecture courses and the MLC mastery courses, but a 1982 study found that students who completed the mastery course did significantly better in subsequent courses. As a result, all sections of the courses were moved to the MLC. The arithmetic and basic algebra courses were removed from the main campus in 1987 as a result of the introduction of institutional admission requirements. The MLC continues to teach three precalculus courses: Intermediate Algebra (2,700 students in 1993), College Algebra (900 students in 1993), and Trigonometry (600 students in 1993). Physically, the MLC consists of six classrooms, a testing center, and a tutoring center.

Course Structure
Each course is divided into a number of units. The unit is most often a single chapter in the textbook, but it may be a combination of chapters. Intermediate algebra has nine units; college algebra has eight units; and trigonometry has five units. Students are graded almost entirely on their performance on the unit tests. These tests are computer generated with each question randomly selected from a question bank containing ten to thirty versions of the question, a method that allows students to retake similar, but not identical, tests. Each test has twenty-one questions and a possible score of 105%. A student must repeat a test until a score of 80% is reached in the mastery class or until a pre-announced deadline has passed in the lecture class. The tests are untimed, are taken outside of class time in a general testing center, and are graded on a strict right-wrong basis. All of the test questions are open-ended rather than multiple choice, and the student's work, as well as the answer, is graded. Only one test may be taken each day, but a test may be reviewed and retaken on the following day. Sixty percent of a student's grade is the average of these test scores plus an optional instructor determined score. The other forty percent is a comprehensive final exam that can be taken only once, unless a student does so poorly that she or he earns a failing grade in the course. Under these circumstances, the student can retake the final exam once for a maximum course grade of C. A minimum score of 60% is required on the final exam in order to earn a C or better in the course.

Originally, all test questions were skills based, but recently we have begun to develop writing questions that are more concept oriented. Students are required to use a graphing calculator in the college algebra and trigonometry courses, and the required use of the graphing calculator is being piloted in intermediate algebra. The fact that all students are allowed to use a graphing calculator in the testing center has challenged us to develop questions oriented more toward student exploration, discovery, and analytical thinking.

Freeing test taking from the confines of the classroom gives the students some distinct advantages: by choosing their own testing times, students can maximize their chance for success; having no time limit keeps momentary panic on a problem from being disastrous; and the fear of failure is minimized since the worst penalty possible is having to retake the exam. Class time is freed up for student interaction. Another advantage cited by MLC students is that they are no longer in direct competition with their classmates--how well or how poorly the student next to them does in no way influences an individual's grade.

Mastery Sections
Mastery classes are scheduled in the same manner as a traditional lecture section with a qualified instructor and four undergraduate tutors assigned to each section of forty-five students. Each student is assigned to a tutor; the instructor monitors the entire class. An advantage of this system is that the student has two people available for help, and generally he or she will respond well to at least one of them. Students are given a list of objectives and homework problems, and may also receive supplementary material, study questions, and sample tests. The student works through the material at her or his own pace, but a recommended assignment schedule is provided.

When a student has completed all of the assigned homework, he or she demonstrates knowledge of the material to the tutor or instructor and is given a "ticket" to take the test on the material. The student takes this ticket to the central testing center and is issued the appropriate preprinted test. Students must earn at least an 80% on each unit test before proceeding to a higher numbered test. When a retake is necessary, the student does some additional problems before receiving a new ticket (referral problems from the textbook are listed at the end of each quiz.) A student may take a test as many times as necessary in order to pass the unit, but instructors generally intervene when a student has three or more tries on a test.

Absolute deadlines are incompatible with the mastery idea, but courses have recommended deadlines. Originally, a ceiling was put on the highest score that a student could earn if he fell behind these deadlines. Unfortunately, the deadline penalties appeared to discourage the struggling student without encouraging the able student, and they were eventually discarded. Currently, students who work ahead of their deadlines are permitted to retake the test for a higher score even if they have previously made an 80%. Students who are behind deadline must accept their first passing score. Two "amnesty" tickets are given to the student when he or she completes the last unit test; the student may use these to retake two tests in order to raise his or her test average, as well as review for the final exam.

The MLC design attempted to build in as much structure as it could, while maintaining the flexibility required by the mastery. Since students must do mathematics on a regular basis in order to learn, attendance is taken, and instructors are encouraged to contact missing students. Instructors file absence reports, and students with excessive absences or no course progress are administratively dropped from the class. A student who is administratively dropped can rejoin the course by fulfilling a "reinstatement agreement" that requires the student to attend regularly and pass an appropriate number of quizzes.

Students have a wide range of entering skill levels, and consequently it is expected that some will require more than one semester to complete a course. In the MLC, mastery students who have good attendance records can earn the RR (reregister) grade by passing approximately 60% of the course contents. These students can register during the next regular semester and complete the course from the previous semester's ending point. Alternatively, some students can finish course material quickly and are allowed to move directly to the next course through an agreement called a "contract." Work done on contract, possibly an entire course, is saved for the student's later enrollment in the course.

Lecture Sections
An attempt was made to develop a lecture course that included the mastery requirement, but students tended to "fall behind" the lecturer, and the course was never very successful. In the current lecture format, students have absolute deadlines after which they may no longer take a test, forcing them to keep up with the lecturer. They are allowed to switch to the mastery section only during a short trial period at the beginning of the semester, and there is no RR option. One undergraduate tutor is assigned to the class for every twelve to fifteen students. Tutors take attendance and grade homework during the lecture, but are available for individual tutoring at the instructor's discretion. Besides the established quizzes, most lecture instructors choose to include an additional instructor-determined component such as in-class quizzes, point incentives for attendance and homework completion, log sheets over textbook reading assignments, or short weekly writing assignments. Standards for test taking (multiple takes permitted) and grading (no partial credit) remain the same for both lecture and mastery sections. This gives the students an instructional choice while equalizing the quality of the preparation of the two groups.

Tutoring Center
Students who are working on their own in a textbook require more help than they can get in three class hours a week. The MLC testing and tutoring centers are currently open seventy-two hours a week. Tutors are scheduled according to demand, with two to twelve available at a given time. Students who are attending class regularly are allowed to review previously taken tests, turn in homework, and receive a ticket for a retake outside class time. Videotapes, provided by the textbook publishers, can be checked out for viewing in the tutoring center or in the home, and the use of the videotapes is steady. Tutorial software, keyed to the textbook, is available in the tutoring center, but student use of the software is light. All of the supplements are optional, and the primary modes of instruction remain the textbook and the instructor or tutor.

Personnel
Instruction in the Learning Center program is done by ten full time nontenure-track instructors and twelve to fifteen graduate assistants. The

nontenure-track instructors have a minimum of a Master's degree, primarily in mathematics or mathematics education, and have full responsibility for curriculum, test development, and the supervision and training of undergraduate and graduate staff. Since no exam writing or grading is required of the instructors, a full-time instructor will typically have six classes. In addition, all instructors have scheduled time in the Tutoring Center, providing help both to their own students and to other students enrolled in the program. An instructor with fewer classes may have fifteen to twenty hours assigned to supervising the activities of the Testing and Tutoring Centers, training and supervising undergraduate tutors, or monitoring of the revising and printing of quizzes and quiz reviews. The importance of having a teaching and tutoring staff who *want to teach at this level* cannot be overstated. The alternative is too often the rotation of these assignments through a reluctant mathematics faculty, most of whom consider such an assignment degrading.

Mathematics graduate students also grade, tutor, and instruct in the program, but since all graduate students must perform satisfactorily in the MLC before receiving another assignment, they are given strong motivation to do well. Graduate students without prior teaching experience or adequate English skills are "stepped through" the jobs of grader, general learning center tutor, classroom tutor, mastery-based instructor, and lecture instructor. New G.A.'s are given a two day orientation to the Mathematics Learning Center before classes begin. A comprehensive instructor manual details the rules and procedures and provides tutoring guidelines. Each graduate assistance is also assigned to an experienced college faculty member as a "teaching partner." Teaching partner assignments are made with the individual schedules in mind, so that G.A. and faculty share times in the Tutoring Center, simplifying communication. G.A.'s are encouraged to use the faculty member as a resource for developing effective classroom strategies and dealing with individual student difficulties. The faculty member is required to visit the G.A.'s class and G.A.'s are encouraged to visit the faculty member's class.

The backbone of the system is the peer tutor. Undergraduate employees have a minimum of one semester of college calculus with a grade of B or better and are primarily math majors, math education majors, and engineering majors. A pay scale with semester and yearly increments encourages tutors to continue with the program, and most students will tutor three to five years with the Mathematics Learning Center. Several monetary awards are given to outstanding tutors each semester, and a social event is usually held each semester. The regular hours and the rewarding nature of the work have made tutoring a popular job, and there are always many applicants. Tutors may also earn credit for tutoring by enrolling in a field experience course offered through the College of Education.

Tutors attend a tutor training meeting before the semester to learn the mechanics of working in the Learning Center. A tutor manual offers both instructions for effective tutoring and specific guidelines for such activities as taking attendance, checking off homework, and recording quizzes. Additional one-hour tutor training sessions are offered at three equally spaced points during the semester. These sessions, offered at three different times over three days to ensure that all tutors can attend, concentrate on developing tutoring skills and familiarity with the subject matter.

Student evaluations of the instructors, tutors, and overall program are done each semester. Tutors receive a summary sheet of responses; instructors may

152

see the actual evaluations after the close of the semester. Instructors evaluate the tutors working with them, and tutors, in turn, evaluate the instructors. A summary of student comments on the program is compiled and posted, and many changes have been made in response to student suggestions on the evaluations.

Computer equipment
Although good personnel is paramount, the program also depends heavily on the computer for its success. Formerly, all computing was done on the campus mainframe (IBM ES/9000), but a new PC-based test generation system has been developed. Development of a new PC-based record keeping system is also underway. Printing is still done on the campus high speed laser printer (IBM 3800). All software for record keeping and test generation has been written by members of the department. The computers make individualized deadlines and placement within courses possible, as well as being used for test generation. Test banks are continuously being revised and expanded as old questions are revised and new questions written.

Summary
The flexibility of the MLC offers many benefits. It is possible for motivated students to complete more than one course in a semester. Students who are not able to complete an entire course within a semester are allowed to take two semesters without penalizing their G.P.A., an option that is not unreasonable considering that these same courses are taught as two-semester courses at the high school level. Because deadlines are set for each individual student, a student who formerly would have dropped a course for which she or he was unprepared (and therefore lost a semester's work in mathematics) is now allowed to drop back into a mastery-based course after the start of the semester. Deadlines are easily adjusted for students with illness or emergencies. Students with physical or learning disabilities and students with extreme math anxiety may choose to work in a small (12-student) special section with a separate support lab for daily instruction. Instructors are notified if students are repeating a course, and close ties are maintained with the ethnic programs.

A math lab provides an efficient use of resources, standardization of course material, uniform performance expectations, and the flexibility for a diverse population of students. It requires a thoughtful and workable initial design, and it must be able to evolve with the student population that it serves. Support from the administration is essential if a nontraditional program is to survive. The most critical ingredient, however, is committed and caring personnel, and every incentive should be given to attract and retain good people. Instructors must have input into policy and curriculum decisions, even though it means some compromises will be made and decisions prolonged. Continuity at every level is vital for the program to mature, and it is no accident that successful programs often have at least one highly committed individual. Developing a math lab will never be the simplest solution for teaching underprepared students, but it may well be a successful one.

PIERCE COLLEGE

MATH LAB

Realizing that all students do not perform well or learn in the standard lecture formatted class, Pierce college has developed a math lab which will meet the needs of some of these students in our college community. The lab, it is hoped, will meet the needs of several groups of students: 1) allowing students who do not have a fixed schedule to take a math class; 2) those who will need additional time to cover the required material; 3) enable those who need a quick review to accomplish that and then move on to another course; 4) allow for a one to one relationship between teacher and student which cannot be accommodated in a standard lecture format.

We have developed a lab which is open 54 hours per week, with day, evening and Saturday hours, thus allowing students to work around their schedule.

There are three basic classes taught in the math lab--fundamental arithmetic, elementary algebra, and intermediate algebra. On occasion, precalculus, trigonometry, and statistics students are accommodated in the lab setting under special arrangements.

The lab is staffed by a faculty member, a lab technician, and one student tutor. (Student tutors are second-year math students who have completed part or all the calculus series.) During the open hours of the lab there are at least two people staffing it. Having people who believe in the lab concept and the ability to make people comfortable is a necessity for a successful lab.

We have a diverse group of students. There are those who have just completed high school and are planning to complete a four-year program. Some of our students need a refresher to get into a vocational or academic program; we have another group who have lost jobs and are coming back to retrain themselves in another field; then we have those who are taking classes for self-enrichment.

When a student enters the college s/he is given a placement test for math, English, and reading. The results of these tests will tell the student in which courses to start their studies.

If they are placed in arithmetic, introductory algebra, or intermediate algebra, they are given the option to take the course in the traditional lecture format or the math lab approach. The students make their decision after being advised by a faculty advisor.

There have been a few problems we've encountered with our math lab. These include: 1) not being able to answer the questions brought by students in an expedient manner; 2) keeping students on task so that they will complete their course; 3) having students do enough practice before trying to take an exam on the material; 4) having students work consistently on their material.

We have tried to incorporate several methods in order to alleviate some of the problems listed above. The following is a typical syllabus given to each student on the first day of class.

INTERMEDIATE ALGEBRA (MATH 101)

5 CREDITS

Instructors: Randy Leifson and Debbie Falcioni

Math Lab Hours: 8:00 am - 4:00 pm, Monday through Friday <u>and</u>
                  5:30 pm - 7:40 pm, Monday through Thursday
                  9:00 am - 2:00 pm, Saturday

GENERAL INFORMATION

This course is intended for students who have just finished a course in Introductory Algebra, and also for students who have studied algebra at some earlier time but still need a course that starts "at the beginning." It is designed to be a self-study tutorial course. You will be taking most of the responsibility for your own learning. Your class will be spent in reading the explanations in the book, working problems, and taking tests. An instructor will always be present to help you when you have problems or questions. In addition, other aids are available to assist you. There is a solution booklet that gives detailed solutions to some of the problems in your text, as well as audio and video tapes that go along with the text. There will also be a daily half-hour lecture help session: the time and topic to be covered are on a poster on the front wall.

The course is also offered in the traditional classroom format.

Your success in this class is up to you. Good luck!

HOW TO PROCEED

You are required to attend for at least five (5) hours a week, when you come to the lab to work on your course material. You should attend during the hour you registered for, unless special arrangements have been made with the instructor. The text book can be purchased at the college bookstore.

HOW TO WORK IN THE TEXTBOOK

1. Each chapter is broken up into sections and each section is divided into several objectives. Read the first objective.

2. Read the explanation of the material and work through the example problems. Then work all the marginal exercises in that section on a separate sheet of paper. (These will be handed in for credit upon completing the chapter.)

3. After completing the entire section, work enough of the additional exercises so that you feel comfortable with the material. (Answers for odd

problems can be found in the back of the book, and the problems worked in detail are in the solution manual available in the lab.)

4. After thoroughly understanding the exercises and the problems in the section, go on to the next section and follow the same procedure.

5. When you have completed all the required sections of the chapter, take the practice test at the end of the chapter. Check it yourself and review the section where you had difficulty. (The test will be handed in for credit.)

6. When you are ready for the test, ask the instructor or aide for the chapter test you need. Allow plenty of time to take the test.

THE QUESTIONS ON THE TEST YOU TAKE WILL BE MODELED ON THE PRACTICE TEST QUESTIONS IN YOUR TEXT.

You should not expect to complete your work entirely during your scheduled hours in the lab. You will need to do homework regularly (up to two (2) hours per day). Don't get behind!

EACH TEST MUST BE COMPLETED IN ONE SITTING. MAKE SURE YOU HAVE ENOUGH TIME TO COMPLETE THE TEST!!!!!

TIME SCHEDULE FOR TESTS

MATH 101

TEST	MATERIAL TO BE STUDIED	NO. OF DAYS	DEADLINE	MASTERY GRADE	SCORE
1.	Chapter 1	5		70	
2.	Chapter 2	6		70	
3.	Chapter 3	7		70	
4.	Chapter 4	6		70	
5.	Chapter 5	6		70	
6.	Chapter 6 (6.1 – 6.4)	5		70	
7.	Chapter 7	5		70	
8.	Chapter 8 (8.1 – 8.3)	5		70	
9.	Chapter 10 (10.1 – 10.5) (Calculator)	5		70	
FINAL _____	To be finished by the last day of class or not more than 8 days after the last chapter test if you finish early.				

158

GRADING SYSTEM

TOTAL POINTS	FINAL DECIMAL GRADE	
1145 – 1077	4.0	95%
1076 – 1066	3.9	
1065 – 1054	3.8	
1053 – 1043	3.7	
1042 – 1031	3.6	
1030 – 1020	3.5	90%
1019 – 1009	3.4	
1008 – 997	3.3	
996 – 986	3.2	
985 – 974	3.1	
973 – 963	3.0	85%
962 – 951	2.9	
950 – 940	2.8	
939 – 928	2.7	
927 – 917	2.6	
916 – 906	2.5	80%
905 – 894	2.4	
893 – 883	2.3	
882 – 871	2.2	
870 – 860	2.1	
859 – 801	2.0	75%

Upon completion of all tests, you can total all your test scores
to determine your grade. All tests must be completed with at
least the mastery grade in order to receive a passing grade.

TESTING AND GRADING

There will be nine 100 point tests and a 200 point comprehensive final. The tests given are like the chapter tests in the book. A more detailed test guide is attached to this syllabus. After each test in the syllabus is listed a number of days and a percentage. The number of days is the number of classroom days allowed for you to pass the test. The percentage is the pass mark or MASTERY GRADE. If you take and pass a test within the allowed time, getting a 90%, that will be the score for you for that test. If you do not pass the test, you must retake it until you obtain the mastery grade or higher. The score entered for you will be either 90% of the score (if taken before your deadline date) or the mastery grade (if taken after your deadline date). You may study longer than the recommended number of days for a chapter and still take the test, but the highest score you will receive will be the mastery grade.

You will also receive 5 points per chapter for completed homework. The homework will consist of the marginal problems and the practice test worked out, showing all steps (to be handed in on your own paper before you take the test for the chapter).

It is important that you understand what your mistakes were on your test. For this reason, we require that you come in to see your test after it has been graded. Your test will always be graded the day after you take it. You can find out your test results by picking up your test slip found in the test slip box on the front desk. (These slips are for your records.)

IMPORTANT LAB POLICIES

1. Any person caught cheating on any exam will receive a 0.0 grade.

2. A person who signs into the lab must stay in the lab for that time. Anyone who is caught signing in and leaving will not be eligible for an "I" grade.

MATH LAB GRADING POLICY

1. All students who complete the required material will receive a grade of "2.0" to "4.0" if:

   a) Complete all chapter tests with at least 70% correct
   b) Complete final with at least 60% correct

2. A student not completing his/her course will receive a grade of "I" (incomplete) if he/she satisfies the following conditions:

   a) Has attended 80% of the classroom sessions
   b) Has completed through the Chapter 6 test successfully

   (If a grade of "I" is earned, the student must complete the course work during the following quarter in order for the "I" grade to be changed to a passing grade.)

3. A student will receive a grade of "0.0" if:

   a) he/she does not complete the course and doesn't satisfy the conditions for an "I" grade
   b) he/she receives an "I" grade and fails to complete the course the following quarter.

ATTENDANCE POLICY

1. Each student is expected to attend one hour a day (5 hours a week).

2. Attendance is taken by each student signing in and out on their attendance sheet.

3. It is the student's responsibility to sign in and out. If a student forgets to sign in, he/she will be considered absent.

4. A student is welcome to spend as much time as he/she wants in lab (over the minimum requirement) as long as there is room.

I have read the entire student handout and understand the policies of the Math Lab. (6 pages)

signature_____

date_____

To meet the needs of our students, we first have each register for a particular hour in which they will attend the lab. There is a class size limit of 20 students per hour, a combination of students from all three courses.

Below is a sample course offering out of one of our class schedule catalogs.

MATH 051 Math Lab Sections:
Orientation the first day of class for each section. Register for one section only. Optional supplemental mini lecture for all Math Lab 051 students from 11:30-12N, M-F.

2900	A	(5)	515	Daily	8-8:50 am	Falcioni
2901	B	(5)	515	Daily	9-9:50 am	Falcioni
2902	C	(5)	515	Daily	10-10:50 am	Falcioni
2903	D	(5)	515	Daily	11-11:50 am	Falcioni
2904	E	(5)	515	Daily	12-12:50 pm	Falcioni
2905	F	(5)	515	Daily	1-1:50 pm	Falcioni
2906	G	(5)	515	Daily	2-2:50 pm	Falcioni
2907	H	(5)	515	Daily	3-3:50 pm	Falcioni
2908	I	(5)	515	MW	5:30-7:40 pm	Leifson
2909	J	(5)	515	TTh	5:30-7:40 pm	Leifson
2910	K	(5)	515	Sat	9-2 pm	Staff

MATH 060 Math Lab Sections:
Orientation the first day of class for each section. Register for one section only. Optional supplemental mini lecture for all Math Lab 060 students, 12-12:30 pm, M-F.

2920	A	(5)	515	Daily	8-8:50 am	Falcioni
2921	B	(5)	515	Daily	9-9:50 am	Falcioni
2922	C	(5)	515	Daily	10-10:50 am	Falcioni
2923	D	(5)	515	Daily	11-11:50 am	Falcioni
2924	E	(5)	515	Daily	12-12:50 pm	Falcioni

(More sections of MATH 060 on next page)

(MATH 060, Math Lab sections, continued)

2925	F	(5)	515	Daily	1-1:50 pm	Falcioni
2926	G	(5)	515	Daily	2-2:50 pm	Falcioni
2927	H	(5)	515	Daily	3-3:50 pm	Falcioni
2928	I	(5)	515	MW	5:30-7:40 pm	Leifson
2929	J	(5)	515	TTh	5:30-7:40 pm	Leifson
2930	K	(5)	515	Sat	9-2 pm	Staff

MATH 101 Math Lab Sections:
Orientation the first day of class for each section. Register for one section only. Optional supplemental mini lecture for all Math Lab 101 students, 12-1 pm, M-F.

2940	A	(5)	515	Daily	8-8:50 am	Falcioni
2941	B	(5)	515	Daily	9-9:50 am	Falcioni
2942	C	(5)	515	Daily	10-10:50 am	Falcioni
2943	D	(5)	515	Daily	11-11:50 am	Falcioni
2944	E	(5)	515	Daily	12-12:50 pm	Falcioni
2945	F	(5)	515	Daily	1-1:50 pm	Falcioni
2946	G	(5)	515	Daily	2-2:50 pm	Falcioni
2947	H	(5)	515	Daily	3-3:50 pm	Falcioni
2948	I	(5)	515	MW	5:30-7:40 pm	Leifson
2949	J	(5)	515	TTh	5:30-7:40 pm	Leifson
2950	K	(5)	515	Sat	9-2 pm	Staff

This method of registration has helped spread our student load throughout the day. We have also used several types of supplementary material, which help take the load off direct contact with an instructor or aide. These include:

1) Audio tapes (by sections of our text)

2) Video tapes (by sections of our text)

3) Computer tutor (we use both Apple and IBM compatibles)

4) Mini-lecture (1/2 hour optional daily lecture for each course)

The student is not restricted only to his/her registered hour. They may come in additional times as long as there is space available.

For those students whose schedules will not fit into a particular registration time, we make special arrangements to meet their needs.

To keep students on schedule, we have incorporated several techniques into our lab:

1) Deadline dates (a set date by which a particular test would have be completed)

2) Homework (this helps ensure each student has covered the material required)

3) A requirement of 5 hours of class attendance each week

4) Mini-lectures which cover the material of each course to be completed by the end of each quarter

For those who start to fall behind in their course work, we try to make personal contacts to see what can be done to eliminate any problems they are having.

Besides teaching the course mentioned above, we also provide tutoring services for students in other classes. This allows the math lab to be a center for math education on the campus. Students in higher level math classes come in for help and also give help to other students who are taking classes they have already completed.

The math lab is and should be an integral part of any college campus.

The math lab will not be the answer to all students' mathematical needs. The success of each student will be dependent on each student's personal drives.

DIRECTORY OF LAB COORDINATORS

BERVER, KITTY
NEW MEXICO STATE
MATHLAB
BOX 3Z
LAS CRUCES NM 88003

EMAIL KBERVER@NMSU.EDU
FAX
OFFICE PHONE EXT 505 646 3901

ALLEN, A DAVID
RICKS COLLEGE
MATHLAB
REXBURG ID 83440

EMAIL
FAX 208 356 2390
OFFICE PHONE EXT 208 356 1926

LEIFSON, RANDY
PIERCE COLLEGE
MATHLAB
9401 FARWEST DRIVE SW
TACOMA WA 98498-1999

EMAIL
FAX
OFFICE PHONE EXT 206 964 6500 X6734

MAGYAR, ANNETE
SOUTHWESTERN MICHIGAN C
CENTER FOR ACADEMIC ACHIEVEMENT
58900 CHERRY GROVE ROAD
DOWAGIAC MI 49047

EMAIL
FAX
OFFICE PHONE EXT 616 782 5113 X228

BATRA, PHYLLIS
UNIV OF WISCONSIN
MCCUTCHEN HALL 206
WHITEATER-800 WEST MAIN
WHITEWATER WI 53190

EMAIL
FAX 414 472 2794
OFFICE PHONE EXT 414 472 3223